Lecture Notes in Bioengineering

More information about this series at http://www.springer.com/series/11564

R. Zainon
Editor

3rd International Conference on Radiation Safety & Security in Healthcare Services

Proceedings of the Thirs, ICRSSHS, Dewan Budaya USM, Penang, Malaysia

Springer

Editor
R. Zainon
Advanced Medical and Dental Institute
Universiti Sains Malaysia
Kepala Batas, Penang
Malaysia

ISSN 2195-271X ISSN 2195-2728 (electronic)
Lecture Notes in Bioengineering
ISBN 978-981-13-4014-7 ISBN 978-981-10-7859-0 (eBook)
https://doi.org/10.1007/978-981-10-7859-0

Softcover re-print of the Hardcover 1st edition 2018

Printed on acid-free paper

This Springer imprint is published by the registered company Springer Nature Singapore Pte Ltd.
part of Springer Nature
The registered company address is: 152 Beach Road, #21-01/04 Gateway East, Singapore 189721, Singapore

Preface

The 3rd ICoRSSiHS is organised by a group of medical physicists, under the Society of Medical Physicist Ministry of Health Malaysia (PERFEKS) since 2012. The conference is designed to provide continuous professional development in the field of radiation safety and security. In each conference, PERFEKS invites IAEA expert to update everyone on the tremendous issues in the field. The local speakers are also invited to share their knowledge and experience in the respected field.

This year, for the first time, ICoRSSiHS is held in Penang, outside the Klang Valley. This conference provides opportunities to all medical physicists involved in radiation safety and security in healthcare service with wide range of activities associated with medical radiation in Malaysia. The main objective is to update medical physicists about the importance of radiation safety and radiation protection to the workers exposed to ionising radiation in the healthcare service. Moreover, this conference is organised to strengthen radiation safety in healthcare services specifically for diagnostic and therapeutic purposes.

This conference is also a platform to the participants to present their results and findings with other participants, and this will motivate them to continue producing high-quality research, study, and innovation. This conference is a platform to provide knowledge, and experience sharing among medical physicists. With the presence of International Atomic Energy Agency speaker, participants can share and updates their knowledge related to medical imaging issues.

The increasing need for irradiating energy at this era is an important factor in the medical and health application in Malaysia and worldwide. It is our obligation to ensure its safe usage at all time. There are three elements related to safety and security of radiation in healthcare services; technological factors, socio-economic factors, and political acceptance. Technological factor is the reason for the power of radiation, and it is in our field of expertise. All of us have to enforce our framework in-line with regulatory requirements, work standards, to give more confidence to the public, working staffs, and decision-makers.

We would like to thank Society of Medical Physicist Ministry of Health Malaysia (PERFEKS), and fellow medical physicists from Penang State Health Department, Perak State Health Department, Kedah State Health Department, Perlis State Health Department, and Universiti Sains Malaysia, who gave full cooperation and support throughout the year to make this event happen. Special thanks to all our sponsors who involved either directly or indirectly in making this conference possible. Thank you to all in making this conference a successful one.

Kepala Batas, Malaysia Organising Committee

Contents

The Influence of Pitch Factor in Reducing Computed Tomography Head Dose Exposure: Single–Centre Trials

N. E. Ismail, F. Mahmood and M. K. A. Karim

Abstract Computed tomography (CT) is mainly associated with high radiation dose exposure to patient and potential for increased risk of cancer. The increasing number of CT head examinations worldwide shows the need for optimization and strategy for dose reduction technique. Thus, the aim of this study is to evaluate the influence of pitch factor in reducing CT dose particularly CT head examination. In this study, the scan acquisition parameter and the CT dose parametric information were collected retrospectively from 16-slice CT scanner (Toshiba Activion) console display. Effective dose (E) was estimated using effective dose per dose-length product (E/DLP) conversion factor, $k = 0.0021$ mSv/mGy.cm. This experiment involved two sets of study, the pre-intervention (n = 163, 58 ± 18 years, 120 kV, 200 mAs, pitch = 0.688) and post-intervention (n = 165, 57 ± 19 years, 120 kV, 200 mAs, pitch = 0.938) on January 2017 and March 2017, respectively. The mean $CTDI_{vol}$ values recorded for pre- and post-intervention were 70.00 ± 8.84 mGy and 51.30 ± 0.72 mGy, respectively. Generally, the mean E value for pre-intervention and post-intervention were 2.75 ± 0.35 mSv and 2.16 ± 0.17 mSv, respectively. It is interesting to note that by increasing the pitch factor in CT head examination has significantly reduced the CT head dose exposure without adversely affecting image quality. The mean DLP value for post-intervention is 1030.10 mGy.cm and has been set as institutional DRL. In conclusion, it is recommended for radiology personnel especially radiographer and radiologist to be aware certain acquisition parameters i.e. pitch factor that work for the optimization process.

N. E. Ismail (✉) · F. Mahmood
Department of Radiology, Hospital Kajang, 43000 Kajang, Selangor, Malaysia
e-mail: erinaismail@gmail.com

M. K. A. Karim
Department of Radiology, National Cancer Institute, 62250 Putrajaya, Malaysia
e-mail: khalis.karim@gmail.com

R. Zainon (ed.), *3rd International Conference on Radiation Safety & Security in Healthcare Services*, Lecture Notes in Bioengineering,
https://doi.org/10.1007/978-981-10-7859-0_1

1 Introduction

CT has developed diagnostic decision making as it provides fast and accurate three dimensional data as compared to other medical imaging tools, hence allowing better patient management [1]. The rapid increase in CT scan examinations is alarming as it is associated to relatively high radiation dose and potential increased risk of cancer. Although it comprises approximately 17% of all medical imaging procedures, it produces approximately half of the population's medical radiation exposure [2]. The increasing number of CT head examinations worldwide shows the need for optimization and strategy for dose reduction technique. Pitch factor is one of the factors affecting CT dose to the patient in the institutional since helical scanning is used as the standard protocol for routine CT head examination. Pitch factor can be defined as the ratio of table translation per 360° tube rotation relative to the nominal beam width in helical CT. A higher pitch indicates faster table translation, and thus larger volume coverage per unit time. Therefore, a higher pitch results in lower radiation dose to the patient, if other parameters including tube current, and kV are kept the same, at the price of lower image quality.

To our knowledge, currently there is no study done on the effect of pitch factor on CT dose for CT head examination. Thus, the aim of this study is to evaluate the influence of pitch factor in reducing CT dose particularly CT head examination.

2 Methodology

The methodology used in this study can be explained under the subtopic CT system, CT dose quantities and data collection.

2.1 CT System

The study involved 16-slice CT scanner (Toshiba Activion). It is the only CT scanner unit available in the institution which has been installed in 2010. The CT scanner was evaluated and tested for quality assurance and quality control protocols regularly for CT Dose Index volume ($CTDI_{vol}$) and dose-length product (DLP).

2.2 CT Dose Quantities

$CTDI_{vol}$ which is a measure of the energy output administered to a single axial "slice" of a patient, and DLP which is an estimation of the total dose administered over the

Table 1 CT dose parametric setting in this study

	No. of sample	Age (years)	kV	mAs	Pitch factor
Pre-intervention	163	58 ± 18	120	200	0.688
Post-intervention	165	57 ± 19	120	200	0.938

entire scan range (z-axis) were collected retrospectively as displayed in the patient protocol on the console display.

2.3 Data Collection

All examination data for consecutive CT head examination in adult (age > 16 years) performed between January 2017 to April 2017 were extracted from INFINITT PACS system which enable the user to extract data such as patient gender, age, scan protocol, $CTDI_{vol}$ and DLP. The experiment involved two sets of study; pre-intervention and post-intervention. Both studies using fixed 120 kV and 200 mAs. The intervention used in this study was the increment of pitch factor from 0.688 to 0.938 (36%) as shown in Table 1. Imaging quality was found to be acceptable following discussion with the radiologists. In order to decrease beam-hardening artefacts, an overlapping pitch ($p < 1$) is used in helical CT head [3]. CT head examination under trauma protocol or contrast-enhanced were excluded from the study.

Based on Eq. 1, effective dose (E) was estimated using effective dose per dose-length product (E/DLP) conversion factor, $k = 0.0021$ mSv/mGy.cm [4]. Pre- and post-intervention data were analysed using Microsoft Excel 2010. The institutional DRL presented in this study are based on mean value (50th percentile) of the dose spread from all patients.

$$E = DLP \times k \tag{1}$$

3 Results

There were 328 adult CT head examinations that were evaluated from January 2017 to April 2017. Table 2 shows the comparison of mean $CTDI_{vol}$, mean DLP and mean E for pre- and post-intervention. The mean $CTDI_{vol}$ for pre- and post-intervention were 70.00 ± 8.84 mGy and 51.30 ± 0.71 mGy, respectively. Generally, the mean E value for pre-intervention and post-intervention were 2.75 ± 0.35 mSv and 2.16 ± 0.17 mSv, respectively. The mean DLP value for post-intervention is 1030.10 mGy.cm and has been set as institutional DLP as recommended by ICRP Publication 103. The institutional DRL was found not to exceed the national DRL, that is 1050 mGy.cm.

Table 2 Computed tomography radiation dose parametrics at 50th percentile

	Pitch factor	$CTDI_{vol}$ (mGy)	DLP (mGy.cm)	E (mSv)
Pre-intervention	0.688	70.00	1308.00	2.75
Post-intervention	0.938	51.30	1030.00	2.16
	% reduction	27	21	21

4 Discussion

It is interesting to note that by increasing the pitch factor by 36% in CT head examination has significantly reduced the $CTDI_{vol}$ by 27%, reduced the DLP and E by 21% without adversely affecting the image quality. The qualitative assessment of the image quality is difficult to be made as the study involved patients with inconsistent patient positioning and patient's head size. It was understood that by increasing the pitch increases the potential for artefacts due to data insufficiency. In this study, the post-intervention CT head images were diagnostically acceptable after discussion with the radiologists.

The reduction of CT head dose exposure in this study leads to the establishment of institutional DRL, which observed to be lower than Canada and Japan DRL as shown in Table 3. However, based on the definition of DRL given by ICRP which is 'a form of investigation level, applied to an easily measured quantity, usually the absorbed dose in air or tissue-equivalent material at the surface of a simple standard phantom or a representative patient' recommends that DRL should be implemented as a benchmark to help operators for optimisation of radiation doses, rather than dose limit.

A process of continuous audit is recommended to guide the appropriateness of institutional scanning parameters and to avoid unnecessarily high doses being delivered to the patient. It is also important to ensure that similar diagnostic quality images are being produced and the DRL produced are within institutional and national limits.

Table 3 Comparison of institutional DRL with national and international DRL

	$CTDI_{vol}$ (mGy)	DLP (mGy.cm)
Ireland 2012 [5]	58	940
Taiwan 2015 [6]	–	999
Switzerland 2010 [7]	65	1000
Australia 2013 [8]	60	1000
Johor 2016 [9]	63	1015
Institutional 50th percentile	51	1030
Malaysia 2013 [10]	–	1050
Japan 2015 [11]	85	1350

5 Conclusion

Increasing pitch factor in CT head examination was found to reduce the CT head dose by 21%. In conclusion, it is recommended for radiology personnel especially radiographer and radiologist to be aware certain acquisition parameters such as pitch factor that work for the optimization process.

References

1. Hricak H et al (2011) Managing radiation use in medical imaging: a multifaceted challenge. Radiol 258(3):889–905
2. Sodickson A (2013) Strategies for reducing radiation exposure from multidetector computed tomography in the acute care setting. Can Assoc Radiol J 64(2):119–129
3. Kalra MK, Sodickson AD (2015) CT radiation: key concepts for gentle and wise use 1. 1706–1721
4. McCollough CH (2008) AAPM Report No. 96 : the measurement, reporting, and management of radiation dose in CT, Alexandria, VA
5. Foley SJ, Mcentee MF, Rainford LA (2012) Establishment of CT diagnostic reference levels in Ireland. Br J Radiol
6. Lin C, Mok GSP, Tsai M, Tsai W (2015) National survey of radiation dose and image quality in adult CT head scans in. 1–12
7. Treier R, Aroua A, Verdun FR, Samara E, Stuessi A, Trueb PR (2010) Patient doses in CT examinations in Switzerland: implementation of national diagnostic reference levels. Radiat Prot Dosim
8. ARPANSA (2012) National diagnostic reference level fact sheet. Aust Radiat Prot Nucl Saf Agency 1–10
9. Karim MKA et al (2016) Establishment of multi-slice computed tomography (MSCT) reference level in Johor, Malaysia. J Phys Conf Ser 694
10. Malaysian Ministry of Health (2013) Malaysian diagnostic reference levels in medical imaging (radiology)
11. Yonekura Y (2015) Diagnostic reference levels based on latest surveys in Japan

Image Quality Evaluation in Contrast Agents Computed Tomography Imaging

J. Zukhi, D. Yusob, A. A. Tajuddin and R. Zainon

Abstract This main goal of this study was to evaluate image quality in single-energy (SE) and dual-energy (DE) CT imaging with the presence of barium and iodine. A fabricated polymethyl methacrylate abdomen phantom with 32 cm diameter size was used to mimic human abdomen. Two different contrast agents: barium and iodine, were scanned separately. The imaging parameters for SECT were set at tube voltage 80, 120 and 140 kV while the imaging parameters for DECT were set at fixed tube voltage 80/140 kV. Both scan modes were set at the different pitch: 0.6 and 1.0 mm, and the slice thickness was set at 3.0 and 5.0 mm with automatic exposure control for the tube current. The CT images obtained from both scanning were analysed to evaluate the signal-to-noise ratio (SNR). Barium and iodine gave highest SNR of 39.30 and 182.68, respectively, at a tube voltage of 140 kV, a pitch of 1 and a slice thickness of 3 mm for SECT. In DECT mode, the highest SNR for barium and iodine were 36.74 and 112.15 respectively at pitch 1 and slice thickness of 3 mm. There was no significant difference between SNR of barium and iodine obtained with both CT imaging modes with p-values of 0.75 and 0.12, respectively.

1 Introduction

The application of the contrast agent has been widely used in clinical work to improve the image quality. Each contrast agent has a different atomic number which influences the attenuation that occurred during the computed tomography (CT) scanning. Contrast agents with the high atomic number are preferable in CT imaging. This is due to their K-edge energy which is within the effective x-ray energies [1–3]. Moreover, the photon starvation effect from the high atomic number of contrast agent is

J. Zukhi · D. Yusob · R. Zainon (✉)
Oncological and Radiological Sciences Cluster, Advanced Medical and Dental Institute, Universiti Sains Malaysia, Bertam, 13200 Kepala Batas, Pulau Pinang, Malaysia
e-mail: rafidahzainon@usm.my

A. A. Tajuddin
School of Physics, Universiti Sains Malaysia, 11800 Minden, Pulau Pinang, Malaysia

R. Zainon (ed.), *3rd International Conference on Radiation Safety & Security in Healthcare Services*, Lecture Notes in Bioengineering,
https://doi.org/10.1007/978-981-10-7859-0_2

not severe [4]. The widespread use of the contrast agents in the clinical application are barium and iodine with atomic number (Z) of 56 and 53, respectively. Their K-edge energy is 37.4 and 33 keV, respectively, which is within the effective x-ray energy. Thus, barium and iodine have high attenuation at low tube voltage either at 80 or 100 kVp [5–7]. However, more noise produced at low tube voltage which causes low image quality [8–10].

CT tube voltage, slice thickness and pitch are factors that affect the CT image quality [11]. By increasing the tube voltage, the photon numbers will be increased which contribute to high signal-to-noise ratio (SNR). Increasing the slice thickness will increase the number of the photon captures in each voxel; thus, more SNR will be calculated, but it will deduct the CT image resolution. The SNR decreases with the increment of the pitch. Pitch is the table movement per 360° rotation divided by the slice thickness [12].Thus, more noise will be obtained when applying high pitch.

Therefore, this study was performed to evaluate the CT image quality with the presence of contrast agents (barium and iodine) in both SECT and DECT imaging. The effect CT imaging parameters on the CT image quality will be investigated in both scan modes.

2 Materials and Methods

The clinical contrast agents of barium (E-Z-CAT) and iodine (IOMERON 350) were used in this study with concentration of 4.9187% w/v (4.6% w/w) and 350 mg iodine/ml, respectively. A single source with dual detector arrays CT scanner (SOMATOM Definition AS, Siemens) was used to scan both contrast agents for SECT and DECT scanning.

2.1 Preparation of Phantom for CT Imaging

A 32 cm diameter of fabricated polymethyl methacrylate (PMMA) spectral CT abdomen phantom was used in this study to mimic adult abdomen. The barium and iodine were filled in a cylindrical tube inside the water phantom. Optically stimulated luminescence dosimeters (OSLD) were inserted in the middle of each cylindrical tube to measure the radiation dose obtained during both scan modes. Figure 1 shows both contrast agents were scanned separately to investigate the effect of CT imaging parameters on CT image quality with the presence of contrast agent.

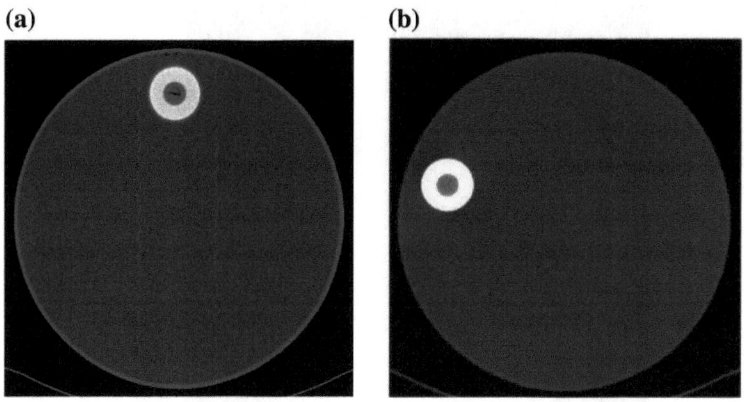

Fig. 1 **a** CT image of barium and **b** iodine in a cylindrical tube in water phantom

2.2 CT Scanning Parameters and Image Analysis

The CT imaging parameters were set at various tube voltages (80, 120, and 140 kV) for single-energy CT (SECT) while 80 and 140 kV was set for dual-energy CT (DECT). Different pitch and slice thickness were applied for both scanning modes at pitch of 0.6 and 1, and slice thickness of 3.0 and 5.0 mm with automatic exposure control (AEC).

Weasis software was used to analyse the CT images obtained from both scan modes and three regions of interest (ROI) were drawn on the CT images to analyse the image quality. The ROIs were drawn within the contrast agent region and water region as the background to get the average mean CT number as shown in Fig. 2. The SNR was calculated as shown in Eq. 1. The S_A, S_B, and σ_B is the mean CT value of contrast agent (HU), mean CT value of background (HU), and standard deviation of background, respectively.

$$SNR = \frac{|S_A - S_B|}{\sigma_B} \tag{1}$$

3 Results and Discussion

The CT images of barium and iodine were analysed in both SECT and DECT modes. The effect CT imaging parameters on the CT image quality was investigated in both scan modes. Analysis CT number of both contrast agents obtained at different tube voltage in SECT mode was shown in Fig. 3. Results show that barium has highest attenuation at low tube voltage (80 kV). Similar profile of CT attenuation was observed at different pitch and slice thickness. This is due to more CT attenuation occurs at lowest tube voltage compared to high tube voltage.

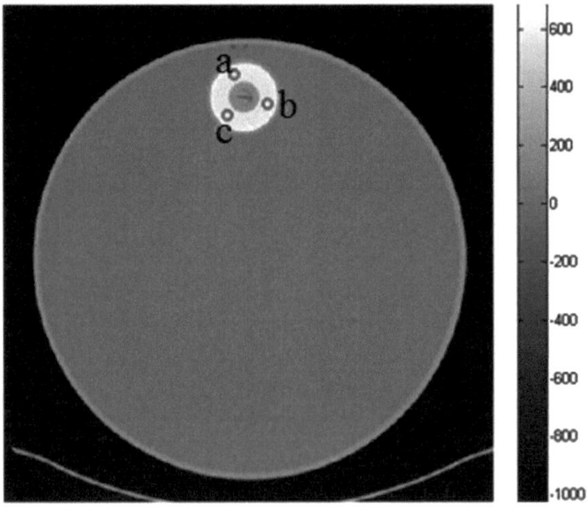

Fig. 2 Three ROIs were drawn on the CT contrast agent image for image quality evaluation

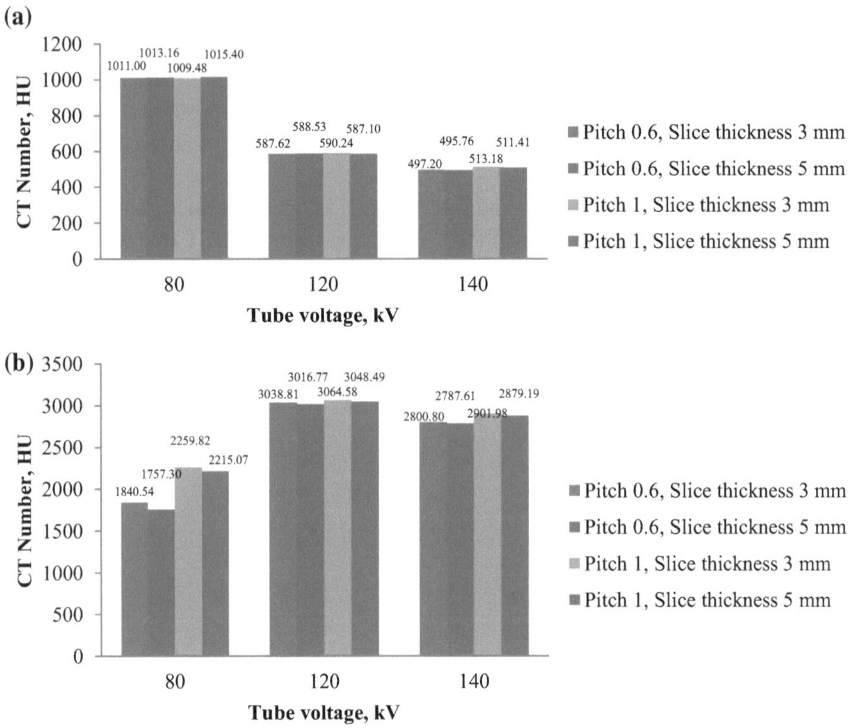

Fig. 3 The CT number of (**a**) barium and (**b**) iodine at different tube voltage in SECT mode

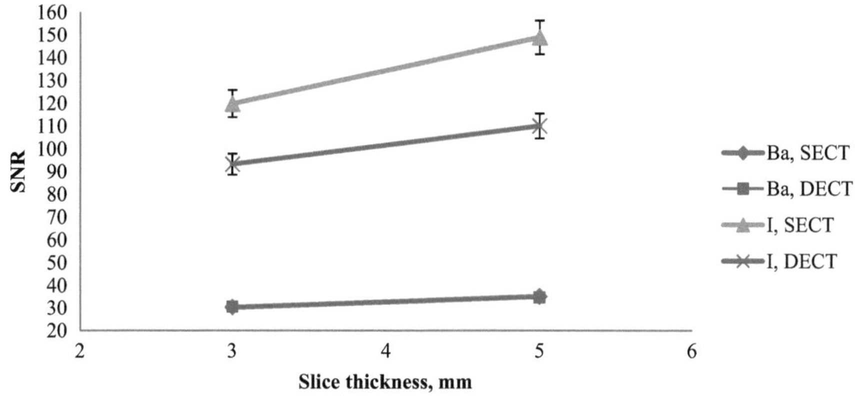

Fig. 4 The SNR of barium and iodine at different slice thickness for both SECT and DECT scan modes at pitch of 1 and tube voltage of 120 kV

The highest attenuation of iodine was observed at 120 kV as shown in Fig. 3b. The K-edge energy of iodine is at 33.0 keV and it is within the low tube voltage energy range. However, iodine gave highest attenuation at 120 kV. This is due to the use of high concentration of iodine in this study that affects the attenuation of iodine. This study found that CT attenuation is also affected by the concentration and the atomic number of contrast agent used in this study. Results obtained in this study is similar to previous studies which found that concentrated iodine gave the highest attenuation at 120 kV [13, 14]. The CT attenuation of iodine gives similar trend at different pitch and slice thickness.

The SNR of both contrast agents at different slice thickness was also investigated. Figures 4 shows the SNR of barium and iodine at different slice thickness from both scan modes at pitch of 1 and tube voltage at 120 kV. Results shows that the SNR increases with the increment of slice thickness. With the use of thicker slices, the number of photons captured in each voxel is high. Thus, more signals were obtained and this factor influences the SNR value for barium and iodine at different slice thickness. Result in this study shows that the SNR of iodine is higher than barium even though its atomic number is smaller than barium. This is due to iodine concentration used in this study is concentrated than barium. The concentration of the contrast agent affects the CT attenuation and, thus, it affects the SNR [15].

This study also found that there was no significant difference in SNR of both contrast agents for SECT and DECT scan mode. This finding is in-line with Zhu and his colleagues, and Yu et al., results where there was no significant difference between SECT and DECT in SNR [16, 17]. The insignificant result is because the SNR of the fused image that obtained from DECT scanning was assumed similar with the 120 kV tube voltage of SECT scanning [18].

Figure 5 shows the SNR of barium and iodine at different pitch value for both scan modes. The SNR decreases with the increment of pitch. Pitch is the table movement per 360° rotation divided by the slice thickness. The image noises obtained

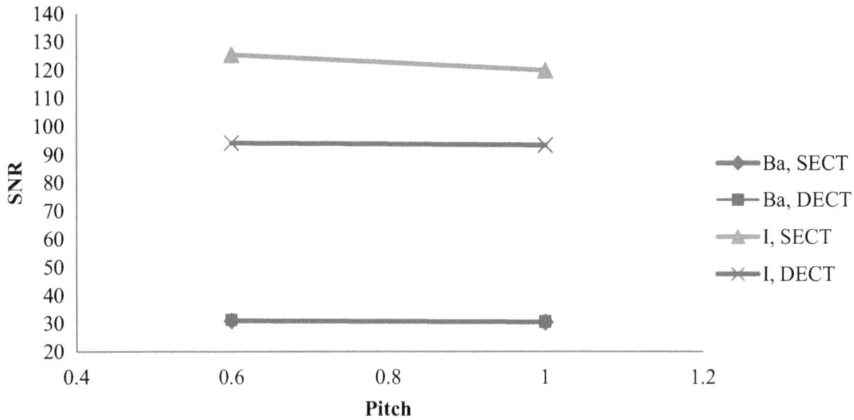

Fig. 5 The SNR of barium and iodine at different pitch for SECT and DECT scan modes

were affected by pitch. Increasing the pitch will cause less projection on the object. Therefore, more noise was obtained on the CT image [12, 19]. On the same note, the SNR is inversely proportional to the noise captured on the CT image. Thus, the higher pitch used gave lower SNR in this study because of more noise were produced. There was no significant difference in SNR of barium for both scan modes at different pitch. However, there was a significant difference in SNR of iodine for both scan modes at different pitch. This is due to the use of concentrated iodine that affect the SNR value.

4 Conclusion

In conclusion, the attenuation of barium and iodine are higher when applying low tube voltage. Barium has higher attenuation at 80 kV tube voltage. However, the use of concentrated iodine makes it more attenuated at higher tube voltage. The CT image quality is affected by a number of factors including the tube voltage, slice thickness, pitch, atomic number and concentration of contrast agent. Increasing the tube voltage and slice thickness will increase the SNR. The SNR decreases with the pitch increment. The SNR obtained from SECT and DECT were not significantly difference for barium when using different slice thickness and pitch. However, iodine had statistically significant between SECT and DECT when applying different pitch but not significantly different when applying different slice thickness. This fundamental knowledge is important to give a better understanding in improving the CT image quality with the presence of multiple contrast agent in CT imaging.

Acknowledgements The authors would like to acknowledge the financial support from Ministry of Higher Education through Fundamental Research Grant Scheme (FRGS).

References

1. He P, Wei B, Feng P, Chen M, Mi D (2013) Material discrimination based on K-edge characteristics. Comput Math Methods Med
2. Anderson NG, Butler AP, Scott NJA, Cook NJ, Butzer JS, Schleich N, Firsching M, Grasset R, Ruiter ND, Campbell M et al (2010) Spectroscopic (multi-energy) CT distinguishes iodine and barium contrast material in MICE. Eur Radiol 20:2126–2134
3. Bateman CJ, Rajendran K, Ruiter NJA, Butler AP, Butler PH, Renaud PF (2015) The hidden K-edge signal in K-edge imaging. 1–7 arXiv:1506.04223v1 [physics.med-ph]
4. He P, Wei B, Cong W, Wang G (2012) Optimization of K-edge imaging with spectral CT. Med Phys 39
5. Bae KT (2010) Intravenous contrast medium administration and scan timing at CT: considerations and approaches. Radiology 256
6. Nakayama Y, Awai K, Funama Y, Hatemura M, Imuta M, Nakaura T, Ryu D, Morishita S, Sultana S, Sato N et al (2005) Abdominal CT with low tube voltage: preliminary observations about radiation dose, contrast enhancement, image quality, and noise. Radiology 237:945–951
7. Buls N, Van Gompel G, Van Cauteren T, Nieboer K, Willekens I, Verfaillie G, Evans P, Macholl S, Newton B, Mey JD (2015) Contrast agent and radiation dose reduction in abdominal CT by a combination of low tube voltage and advanced image reconstruction algorithms. Eur J Radiol 25:1023–1031
8. Fessler J (2009). X-ray imaging : noise and SNR. Chapter 6, X-ray imaging noise SNR. 1–11
9. Hanson KM (1981) Noise and contrast discrimination in computed tomography. Radiol Skull Brain 5:3941–3955
10. Chaudhari A (2012) Improving signal to noise ratio of low-dose CT image using wavelet transform. Int J Comput Sci Eng 4:779–787
11. Goldman LW (2007) Principles of CT: radiation dose and image quality. J Nucl Med Technol 35:213–226
12. Goldman LW (2008) Principles of CT: multislice CT. J Nucl Med Technol 36:57–68
13. Yu L, Liu X, Leng S, Kofler JM, Ramirez-giraldo JC, Qu M, Christner J, Fletcher JG, McCollough CH (2012) Radiation dose reduction in computed tomography: techniques and future perspective. Imaging Med 1:65–84
14. Zukhi J, Yusob D, Tajuddin AA, Vuanghao L, Zainon R (2017) Evaluation of image quality and radiation dose using gold nanoparticles and other clinical contrast agents in dual-energy Computed Tomography (CT): CT abdomen phantom. IOP Conf Ser J Phys Conf Ser 851
15. Bongers MN, Schabel C, Krauss B, Claussen CD, Nikolaou K, Thomas C (2017) Potential of gadolinium as contrast material in second generation dual energy computed tomography—An ex vivo phantom study. Clin Imaging 43:74–79
16. Zhu X, Mccullough WP, Mecca P, Servaes S (2016) Dual-energy compared to single-energy CT in pediatric imaging: a phantom study for DECT clinical guidance. Pediatr Radiol 46:1671–1679
17. Yu L, Primak AN, Liu X, Mccollough CH (2009) Image quality optimization and evaluation of linearly mixed images in dual-source, dual-energy CT. Med Phys 36:1019–1024
18. Siemens AG (2008) Dual energy CT SOMATOM definition
19. Primak AN, Mccollough CH, Michael R, Zhang RTRJ, Fletcher JG (2006) Relationship between noise, dose, and pitch in cardiac multi—detector row CT. Radiographics. 1785–1795

Study on Different Method to Determine the Individual Diameter for Size-Specific Dose Estimates (SSDE) in Adult Patients

N. M. Huzail, M. A. A. M. Roslee, N. S. M. Azlan and N. D. Osman

Abstract The CT dose index (CTDI) and dose length product (DLP) are the most frequently used indicators to represent radiation doses in CT examination. However, the limitation of both is that they only estimate dose based on the scanner output information for specific standardized condition and phantom sizes. This study was aimed in evaluation of the radiation dose based on the SSDE method for adult abdomen CT study at AMDI, Universiti Sains Malaysia, Penang. A total of 91 CT procedures were selected consisting only adult patients undergo CT thorax-abdomen-pelvis (TAP) examination. As recommended by American Association of Physicist in Medicine (AAPM), the individual dimensions of each patient were determined. The conversion factors for SSDE were multiplied with the displayed $CTDI_{vol}$. The comparative study between displayed $CTDI_{vol}$ with SSDE-calculated dose were done and percentage difference were then determined. From the results, significant differences were observed between SSDE-calculated and displayed dose with variations of 3–47% for method of AP and LAT summation, and 1.96–46% with method of effective diameter. The SSDE calculated doses were significantly higher than the displayed dose values by CT scanner. Therefore, evaluation of patient dose by individual specific size is critical for optimization of radiation exposure in CT imaging.

N. M. Huzail
School of Physics, Universiti Sains Malaysia, 11800 Minden, Penang, Malaysia
e-mail: nurhanishuzail@gmail.com

M. A. A. M. Roslee
Imaging Unit, Advanced Medical and Dental Institute, Universiti Sains Malaysia, 13200 Kepala Batas, Penang, Malaysia
e-mail: amirul.azrie@usm.my

N. S. M. Azlan
Faculty of Computer and Mathematical Sciences, Universiti Teknologi MARA, 15050, Kota Bharu, Kelantan, Malaysia
e-mail: noorsyamimimuhamadazlan@gmail.com

N. D. Osman (✉)
Oncological & Radiological Sciences, Advanced Medical and Dental Institute, Universiti Sains Malaysia, Bertam, 13200 Kepala Batas, Penang, Malaysia
e-mail: noordiyana@usm.my

R. Zainon (ed.), *3rd International Conference on Radiation Safety & Security in Healthcare Services*, Lecture Notes in Bioengineering,
https://doi.org/10.1007/978-981-10-7859-0_3

Keywords Computed tomography · Size-specific dose estimation (SSDE)
Effective diameter · CT dose index (CTDI)

1 Introduction

Currently, the volumetric CTDI ($CTDI_{vol}$) and dose length product (DLP) values have been employed to represent radiation doses from CT examination [1]. The $CTDI_{vol}$ is defined for two standard polymethylmethacrylate (PMMA) phantoms; a diameter of 32 cm to represent the patient's body and 16 cm to represent the patient's head [2, 3, 8]. However, both metrics are not considered to represent the actual patient dose but only the measurement of machine output [1–5].

The main controversial issue with current dose metrics is the underestimation of the patient dose by 40–70% for averaged-size adult and paediatric torsos [1, 5, 7]. This problem led to the beginning of new technique known as size-specific dose estimate (SSDE) by American Association of Physicist in Medicine (AAPM). AAPM introduced the conversion factors for an accurate dose estimation as proposed in Report 204 [1–9]. This report combined four independent research teams that have studied the potential of patient size-dependent factors to estimate the true patient dose from the displayed $CTDI_{vol}$ [10]. Four different calculation methods were used to represent the patient size such as the anteroposterior (AP) dimension, the lateral (LAT), the sum of both dimensions (AP + LAT), and the effective diameter (Deff).

In this study, two different approaches to calculate SSDE were compared. The calculated SSDE were derived from two conversion factors; one from the summation of AP + LAT dimensions and another is the effective diameter. Both different methods to determine individual patient's diameter were compared.

2 Method

Institutional committee review for ethical approval on the clinical data study was obtained (USM/JEPeM/17030180). All CT imaging performed at Imaging Unit, AMDI USM, Penang, Malaysia were performed with dual-energy SOMATOM Definition AS + CT scanner (Siemens Healthcare, Germany). This study involved a retrospective survey on selected patient data that only included the patient dose received from CT TAP study. This retrospective survey was carried out on patient data retrieved from the Picture Archiving and Communication System (PACS) between the periods of June 2016 until March 2017.

2.1 Survey and Determination of Individual Diameter for Adult Abdomen

The collected patient data consists of individual patient's information such as age, gender, examination date, CT procedure, $CTDI_{vol}$, DLP, and scan parameters (kVp, mAs, scan length and slice thickness). The individual AP and LAT diameters for each patient were measured at similar anatomic landmark through the relatively largest extent for a reliable result. The measurements were done at three sub regions, which were mid spleen, mid kidney, sacral level by using digital caliper tool on the CT workstation console.

Figure 1 shows the measured LAT and AP projections for three different anatomic scan locations from the CT images. The LAT diameter was measured at the central image as it is the widest region of lateral dimension. However, the measurement of AP diameter was measured at the widest point of AP dimension. The diameter measured for each different location were then compared.

The individual effective diameter (D_{eff}) of each patient was determined using the root of the product of the AP and LAT patient measurements as defined by the equation provided in AAPM Report 204 [5].

$$Effective\,diameter = \sqrt{AP \times LAT} \quad (1)$$

2.2 Calculation of Radiation Dose Based on SSDE

Size-specific dose estimate (SSDE) conversion factors recommended by AAPM were derived from measured patient size. The AP and LAT measurements and the computed effective diameter were used to determine the conversion factor. To determine the actual patient dose, the SSDE was calculated by multiplying the $CTDI_{vol}$ extracted from the PACS with the conversion factors provided by Report 204 [5]. The equation for SSDE-calculated dose is

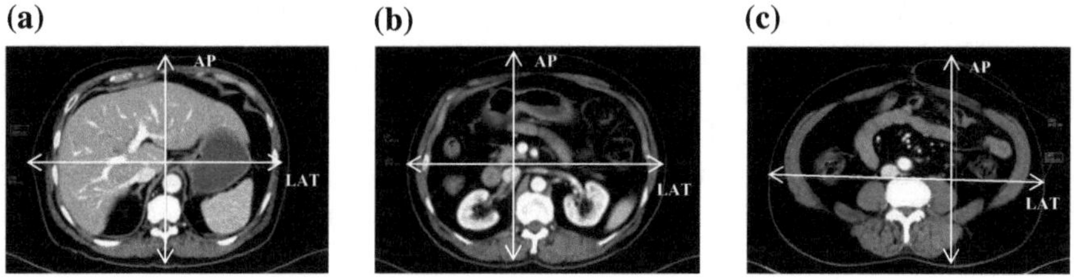

Fig. 1 The measurement of both lateral and AP dimensions at three different subregions: **a** mid spleen, **b** mid kidney and **c** sacral region

$$SSDE = CTDI_{vol}^{32} \times f_{size}^{32} \quad (2)$$

The SSDE-calculated dose derived from two different conversion factors; AP + LAT summation and D_{eff} measurements were compared and the percentage differences were determined.

2.3 Comparative Study of Displayed Dose Versus SSDE-Calculated Dose

The percentage difference (% differences) between the displayed patient dose, $CTDI_{vol}$ and the standard dose from SSDE calculation was determined to compare the displayed dose with the corrected dose based on individual patient's size. The percentage difference was determined using below equation.

$$\%\,Differences = \frac{SSDE - Displayed\,CTDI_{vol}}{SSDE} \times 100 \quad (3)$$

3 Results

3.1 Survey and Determination on Effective Diameter of Adult Abdomen

A total of 91 adult patients registered for CT TAP examinations were included in this study, with 32 males and 59 female patients. The patients' ages were ranged between 23 years and 88 years old and the patient mean age was 53.6 years ± 12.3. The patient data were arranged into 7 age groups based on their age (21–30, 31–40, 41–50, 51–60, 61–70, 71–80, 81–90 years old).

The relationship between patient age and the effective diameter of these three level sub-regions are shown in Fig. 2. From the graph, it can be observed that the the highest effective diameter was at sacral region for most of the aged groups. The mean effective diameter of each group were 25.4 cm at mid spleen level, 25.5 cm for mid kidney level, and 26.2 cm for sacral level.

The effective diameter of these three sub-anatomic regions were then averaged together and defined as mean effective diameter for individual patients, as depicted in Table 1. These mean values of D_{eff} were used to determine the conversion factors for the SSDE-calculated dose. Table 1 shows comparison between the two different methods for determination of patient diameter. The highest dimension from calculated of AP thickness and LAT width (AP + LAT) was found in 37 years old female, while the lowest diameter was observed in 41 years old female patient.

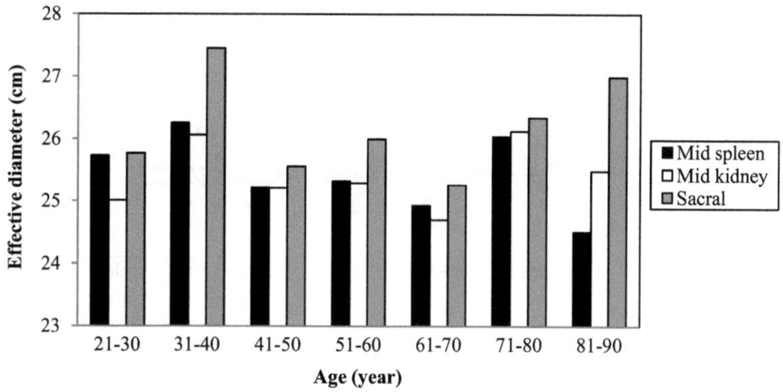

Fig. 2 Relationship between effective diameter of each sub-level with patient age

Table 1 Patient-specific size measurements

Age group (years)	AP + LAT (cm)			D_{eff} (cm)		
	Mean ± SD	Min	Max	Mean ± SD	Min	Max
21–30 (N = 2)	51.96 ± 3.97	49.15	54.76	25.51 ± 2.04	24.06	26.95
31–40 (N = 10)	54.26 ± 8.48	46.45	69.78	26.59 ± 4.24	22.51	30.95
41–50 (N = 18)	51.63 ± 7.47	38.92	69.39	25.41 ± 3.73	18.96	34.50
51–60 (N = 35)	52.03 ± 6.08	42.81	68.51	25.69 ± 3.05	20.93	33.94
61–70 (N = 17)	50.68 ± 5.46	41.59	62.98	24.96 ± 2.74	20.20	31.14
71–80 (N = 7)	53.24 ± 6.30	43.46	63.74	26.17 ± 3.19	21.45	31.72
81–90 (N = 2)	52.09 ± 7.13	47.05	57.13	25.66 ± 3.42	23.24	28.08

Mean overall in the AP + LAT dimension was 52.0 cm ± 6.4, with a range of 38.9–69.8 cm and were corresponded with a range of conversion factor of 1.04 up to 1.87. The mean effective diameter computed from the AP thickness and LAT width for each data series was 25.6 cm ± 3.2 and was ranged from 18.96 to 34.50 cm for overall age groups.

Figure 3 shows the scatterplot that describe the distribution of measured individual diameter using two different methods, which were the summation of AP + LAT (Fig. 3a) and efective diameter (Fig. 3b) for both gender as a function of patient age. From these figures, the slope of the line shows a very weak relationship between the measured diameter and age for both female and male patients. For first method (AP + LAT summation), results showed weak correlation between patient age and measured diameter for both male and female with R^2 value of 0.31% and 0.66%, respectively. For AP + LAT method, the statistical analysis showed significant difference for both gender, with p-value of 0.812 and 0.854, for both male and female respectively.

In Fig. 3b, the scatter plot shows the distribution of measured dose using D_{eff} dimension for both genders as a function of patient age. From the figure, it can be observed that the relationship between patient size, D_{eff} of both male and female

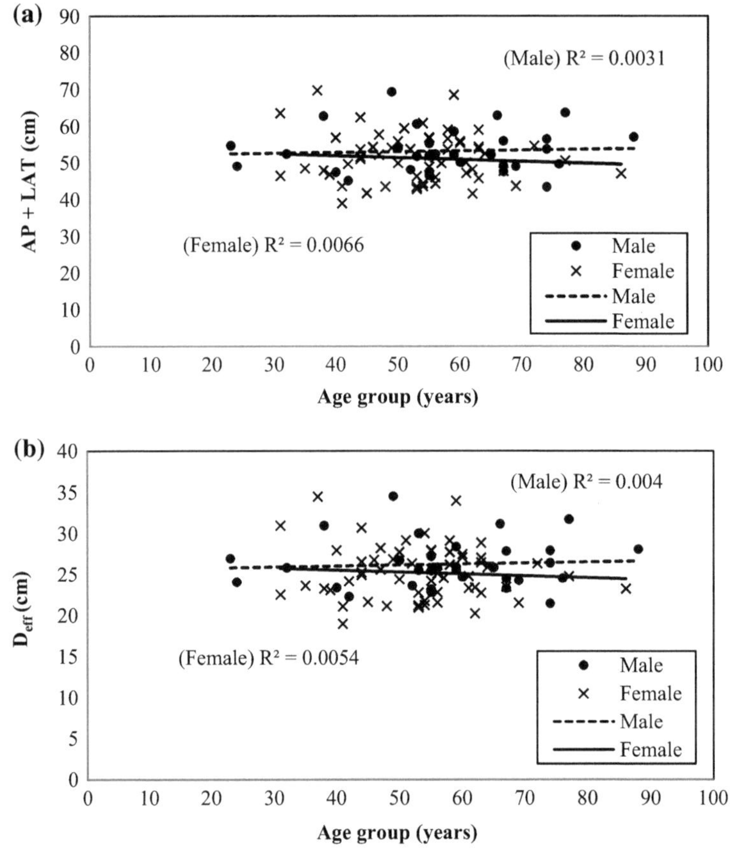

Fig. 3 The individual diameter determined from two different methods; **a** summation of AP + LAT dimension and **b** the effective diameter is shown as a function of age for female and male patient group

patient were very weak, since the value of R^2 is 0.4% and 0.54%, respectively. However, the statistical analysis showed significant difference between gender, with p-value of 0.794 and 0.781, for both male and female respectively.

3.2 *SSDE-Calculated Dose, $CTDI_{vol}$*

Based on the dose measurement among the 91 patients, the mean displayed $CTDI_{vol}$ was 8.6 mGy ± 3.1, which is based on standard phantom size. However, the mean SSDE-calculated dose derived from both AP + LAT and effective diameter calculations resulted in higher dose, which were 11.9 mGy ± 2.8 and 11.8 mGy ± 2.8, respectively.

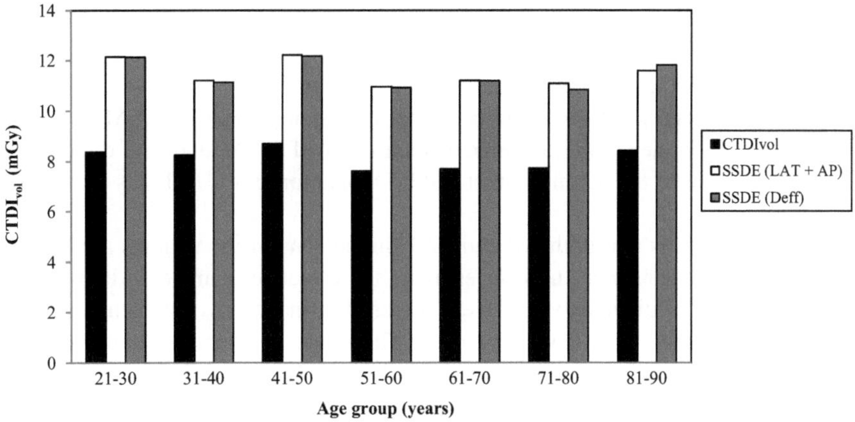

Fig. 4 The comparison of SSDE-calculated dose between two different methods for diameter measurement and displayed dose for each age group

As depicted in Fig. 4, the comparison of SSDE-calculated dose between different method of patient size measurements, which were AP + LAT dimensions and effective diameter as a function of patient age. From the observation, the SSDE-calculated doses derived from AP + LAT method were slightly higher than D_{eff} measurement for most of the age group with small percentage difference of 1–3%. These results denoted that the overall agreement between $SSDE_{(Deff)}$ and $SSDE_{(AP + LAT)}$ calculated $CTDI_{vol}$ for patient age 23–88 years was an average of 1.90%, with a maximum variation of 2.59%.

3.3 *Comparative Study of Displayed $CTDI_{vol}$ Versus SSDE-Calculated $CTDI_{vol}$*

Figure 4 shows comparison of displayed $CTDI_{vol}$ and calculated-SSDE doses for both methods as a function of patient age. For patients aged between 23 and 88 years, the calculated-SSDEs were approximately a factor of 1.46 times greater than the displayed $CTDI^{32}_{vol}$.

Besides, when the source of SSDEs was analysed, overall agreement of no more than 50% were observed between the displayed and calculated dose. The variations were slightly higher for dose derived from summation of AP and lateral dimensions of 3–47%, compared with 1.96–46% when effective diameter was applied.

4 Discussion

SSDE conversion factors can be determined for a particular patient from the AP thickness and LAT width from the table that has been provided in AAPM Report 204 [5]. Based on the report, the range of summation of the AP and LAT dimensions based on the use of the 32 cm diameter PPMA phantom for $CTDI_{vol}$ is from 16 to 90 cm.

In the conditions that both AP and LAT dimensions of the patients are known, these two dimensions were used to estimate the effective diameter. Alternatively, calculated effective diameter values were also used to find the f_{size} that then multiplied by $CTDI_{vol}$, yields SSDE calculated dose. The range of effective diameter in the Report 204 is limited from 8 to 45 cm. Meanwhile, ICRU Report 74 has provided the option to estimate effective diameter as a function of patient age. The corresponding effective diameter provided in the Report 74 can be further used to determine the conversion factors, f_{size} in AAPM Report 204 [1, 5]. However, using the age only to calculate SSDE were impossible in this study because of a lack of data published in ICRU 74 for patients older than 18 years. In fact, as the Report 204 has provided size-based variable, patients' sizes are at an advantage to find the most appropriate f_{size}.

Figure 4 demonstrated slightly differences between calculated SSDEs that come from AP + LAT or effective diameter calculations due to the drawback of translating three-dimensional object into two-dimensions [1]. These results give justification for the combination of AP and LAT measurements, either as a summation or effective diameter calculation is more preferable for calculating SSDE than either measurement used individually similar to Braudy Kaufman et al suggested previously [1].

From Fig. 4, variability between displayed $CTDI_{vol}$ and calculated SSDEs for a given patient size were observed because of the expected variability in patient size as the reported $CTDI_{vol}$ is known to be based on the specific diameter of 32 cm for body dosimetry phantom. Christner et al reported that the spread out in $CTDI_{vol}$ for individual patient size was noticed even for patients of the same AP + LAT dimension because of the expected variability in patient body habitus and selected scan range [7]. Thus, by multiplying different $CTDI_{vol}$ values with the same f_{size}, the computed SSDE would also differ.

Based on the Report 204, for patient's diameter higher than 72 cm (for AP + LAT) and 35 cm (effective diameter), the f_{size} is smaller than 1. Thus, the calculated SSDE obtained by multiplying the $CTDI_{vol}$ with f_{size} will be smaller than the displayed dose. This subsequently result in enlarge of SSDE for smaller-sized patients. In the evaluation of the data sets studied, a range of patient sizes indicated by the AP + LAT measurements and effective diameter were smaller than 72 cm and 35 cm respectively, thus the SSDE calculated dose derived were expected to be higher than the displayed $CTDI_{vol}$. It is therefore proved that patient size is more important in CT acquisitions performed.

Our study has some limitations. The sample population in this study only included 91 adult patients that undergo TAP CT examinations and it may not the optimal sample size for SSDE calculations. Besides, because of individual habitus is differs, consistent anatomical landmark is also difficult to be measured for each patient. In this study, only patient age was used for correlation with the SSDE-calculated dose. Patient age alone may not be the best variable to correlate with radiation dose.

5 Conclusions

The combination of LAT and AP measurements, as a summation or effective diameter should be practicable to determine SSDE conversion factors to estimate patient dose more accurately. The SSDE calculated doses were significantly higher than the displayed dose values more accurately. Therefore, evaluation of patient dose by individual specific size is critical to optimize the use of radiation exposure in CT imaging and to ensure that dose received by patients is as low as reasonably practicable. Future research should include bigger sample size evaluation including patient weight that will deliver more accurate SSDE calculation.

Acknowledgements The authors would like to thank all radiographers and staff at Imaging Unit, AMDI USM, Malaysia for their help and support throughout this work.

References

1. Brady SL, Kaufman RA (2012) Investigation of american association of physicists in medicine report 204 size-specific dose estimates for paediatric CT implementation. Radiology 265(3):832–840
2. Brink JA, Morin RL (2012) Size-specific dose estimation for CT: how should it be used and what does it mean? Radiology 265(3):666–668
3. Anam C, Haryanto F, Widita R, Arif I, Dougherty G (2016) A fully automated calculation of size-specific dose estimates (SSDE) in thoracic and head CT examinations. J Phys: Conf Ser 694(1):012–030
4. Boone JM, Strauss KJ, Cody DD, McCollough CH, McNitt-Gray MF, Toth TL (2013) Size-specific dose estimates (SSDE) in paediatric and adult body CT examinations: AAPM report no. 204 American Association of Physicists in Medicine website. 1–26
5. McCollough CH, Leng S, Yu L, Cody DD, Boone JM, McNitt-Gray MF (2011) CT dose index and patient dose: They are not the same thing. Radiology 259(2):311–316
6. Christner JA, Braun NN, Jacobsen MC, Carter RE, Kofler JM, McCollough CH (2012) Size-specific dose estimates for adult patients at CT of the torso. Radiology 265(3):841–847
7. Huda W, Vance A (2007) Patient radiation doses from adult and paediatric CT. Am J Roentgenol 188(2):540–546
8. Brady SL, Mirro AE, Moore BM, Kaufman RA (2015) How to appropriately calculate effective dose for CT using either size-specific dose estimates or dose-length product. Am J Roentgenology 204(5):953–958
9. Noferini L, Fulcheri C, Taddeucci A, Bartolini M, Gori C (2014) Considerations on the practical application of the size-specific dose estimation (SSDE) method of AAPM Report 204. Radiol Phys Technol 7(2):296–302

10. Moore BM, Brady SL, Mirro AE, Kaufman RA (2014) Size-specific dose estimate (SSDE) provides a simple method to calculate organ dose for pediatric CT examinations Med Phys 41(7)

Investigation of Patient Dose Received During Digital Dental Radiography and Comparison with International Diagnostic Reference Level (DRL)

F. H. Azhar, H. A. Jaafar, N. S. M. Azlan and N. D. Osman

Abstract Patient exposure from digital dental radiography has increased rapidly and contributes to one of the major sources for radiation doses received in diagnostic radiology. Therefore, assessment on patient dose is crucially important for dose optimisation and justification. However, there are currently no national diagnostic reference level (NDRL) has been established in Malaysia for dental radiography. This study was aimed to evaluate the patient doses received from digital dental radiography which includes orthopantomogram (OPG), cone beam computed tomography (CBCT), and cephalometric examinations at Advanced Medical and Dental Institute (AMDI), Malaysia and serves as preliminary work in establishment of local DRL. The data of dose-area product (DAP) values for all dental examinations performed were gathered started from May 2015 until March 2017. For comparative study, the third quartiles for each examination were determined and compared with the international DRL. A total of 684 cases, consisting of OPG (67%), CBCT (6%), and cephalometric examinations (27%) performed during that period. The DAP values were ranged between 47.4 and 363.2 mGy cm^2 for OPG, 392.4 and 1254.4 mGy

F. H. Azhar
School of Physics, Universiti Sains Malaysia, 11800 Minden, Penang, Malaysia
e-mail: ftn.hanani92@gmail.com

H. A. Jaafar
Imaging Unit, Advanced Medical and Dental Institute, Universiti Sains Malaysia, 13200 Kepala Batas, Penang, Malaysia
e-mail: hanisarina@usm.my

N. S. M. Azlan
Faculty of Computer and Mathematical Sciences, Universiti Teknologi MARA, 15050 Kota Bharu, Kelantan, Malaysia
e-mail: noorsyamimimuhamadazlan@gmail.com

N. D. Osman (✉)
Oncological & Radiological Sciences Cluster, Advanced Medical and Dental Institute, Universiti Sains Malaysia, Bertam, 13200 Kepala Batas, Penang, Malaysia
e-mail: noordiyana@usm.my

R. Zainon (ed.), *3rd International Conference on Radiation Safety & Security in Healthcare Services*, Lecture Notes in Bioengineering,
https://doi.org/10.1007/978-981-10-7859-0_4

cm^2 for CBCT, and 17.4 and 33.3 mGy cm^2 for cephalometric procedure. From the results, it showed that patient dose mainly depends on image acquisition protocol. Patient dose assessment is significant for radiation protection management in clinical practice.

Keywords Digital dental radiography · Dose-area product
Diagnostic reference levels

1 Introduction

Recently, patient exposure to medical and dental X-ray examination has grown rapidly and diagnostic radiology represents the largest source of artificial radiation as compared to natural background exposure [1]. The roles of digital imaging such as panoramic radiographic or OPG, cone beam CT (CBCT), and cephalometric examinations have increased in the dental imaging field; therefore, the patient dose is attracting more attention [2]. As to protect patients from unnecessary radiation, the concept of diagnostic reference levels (DRLs) have been introduced and applied for dose optimization in clinical practice. However, currently there is no national diagnostic reference level (NDRL) has been established in Malaysia for dental radiography.

In 1996, IAEA has proposed a dose guidance level as an optimization strategy for dose reduction in radiological procedures [3] and ICRP has recommended the use of DRL for patients undergo medical exposure [4]. The DRL indicates whether the dose received by patient from a particular imaging procedure is high or low. As suggested, the DRL were established based on the third quartile value for the distributions of doses [5]. These reference levels are expected not to be exceeded for a standard procedure if good practice regarding diagnostic and technical performance is applied [6].

This study was intended to evaluate the patient doses received from digital dental imaging including OPG, CBCT and cephalometric examinations and to compare the local radiation dose with the established international DRLs for dental imaging. The comparative study of local dose distribution in digital dental imaging will help to assess the current technique used in routine clinical practice and to optimise the use of radiation exposure in local practice. This research will also provide a small dose database for the establishment of local DRL and further used as guidance data for Malaysian DRL in digital dental imaging.

2 Methods

Institutional committee review for ethical approval on the clinical data study was obtained (USM/JEPeM/16040164). There were no exclusion criteria for the selection of patients as all the patients' data undergo digital dental imaging were included in this study. All data were treated confidentially. This study was carried out on patient dose received from digital dental imaging carried out by Planmeca Promax 3D MID system (Planmeca Oy, Helsinki, Finland) at Imaging Unit, Advanced Medical and Dental Institute (AMDI), Universiti Sains Malaysia (USM), Penang, Malaysia.

2.1 Survey on Local Patient Dose Data

A retrospective survey was performed on patient dose data received from digital dental imaging including panoramic or OPG, CBCT and cephalometric procedures. The dose data represented as dose-area product (DAP) values were retrieved from Planmeca Romexis system (version 3.7.0.R). The exposure parameters setting such as tube voltage and current, exposure time, and field of view (FOV) and patient information including patient's age, and gender were also collected for all patients. All data were collected started from May 2015 until March 2017.

The statistical analysis was performed on the collected patient dose data of cephalometric, OPG, and CBCT procedure. The minimum, mean, maximum, first quartile, and third quartile values of DAP were calculated respectively for each dental examination which were cephalometric, OPG, and CBCT procedure.

2.2 Comparative Study of Local Dose with the International DRLs

For comparative study, the local dose data were represented by the third quartile values of the collected DAP distribution. The calculated third quartile of DAP values for each procedure were then compared with the selected international DRLs. In this study, the national UK DRL [7, 8] were used as the standard reference for the local dose data for cephalometric and OPG procedures, as shown in Table 1. However, the reference value used for dental CBCT examination was based on achievable dose as recommended by SEDENTEXCT European guidelines [9].

The collected local dose data were documented for local radiation protection program and further clinical audit on the current practice at the Imaging Unit, AMDI USM, Malaysia. The collected data were also documented for establishment of future local DRL.

Table 1 Suggested international DRL in dental radiography

Examination type	DAP per radiograph (mGy cm^2)
[a]Cephalometric	40
[a]Panoramic/OPG	93
[b]Cone beam CT (CBCT)	250

[a]The DAP values are based on the recommended national DRL of UK [7, 8]
[b]Achievable dose as recommended by European guidelines on CBCT for adult [9]

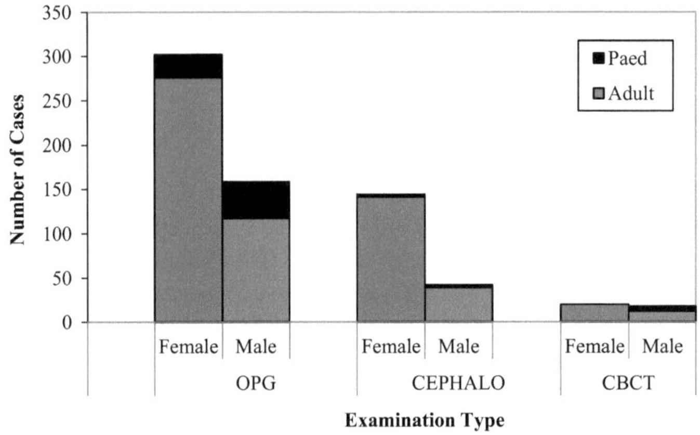

Fig. 1 Number of cases of patients based on gender undergo dental imaging from May 2015 until March 2017

3 Results and Discussion

3.1 *Survey on Local Patient Dose Data*

From the dose survey done, a total of 684 cases of digital dental imaging including cephalometric, OPG and CBCT procedure were performed and collected during the period of survey. The distribution of cases for each dental imaging procedures obtained in this survey representing both gender and age group was depicted in Fig. 1. The highest number of patient that undergo digital dental imaging was OPG examination with 67% of the total cases, followed by cephalometric imaging (27%) and CBCT imaging (6%). In most study, the number of adult patients were higher than paediatric patients which was only 12% of total cases. Besides, the number of female patients undergo digital dental imaging were higher compared to male, which were OPG (66%), CBCT (53%), and cephalometric (77%), respectively.

Figure 3 shows the distribution of collected DAP values based on the patients age for each digital dental imaging. The relationship of DAP with patients age were

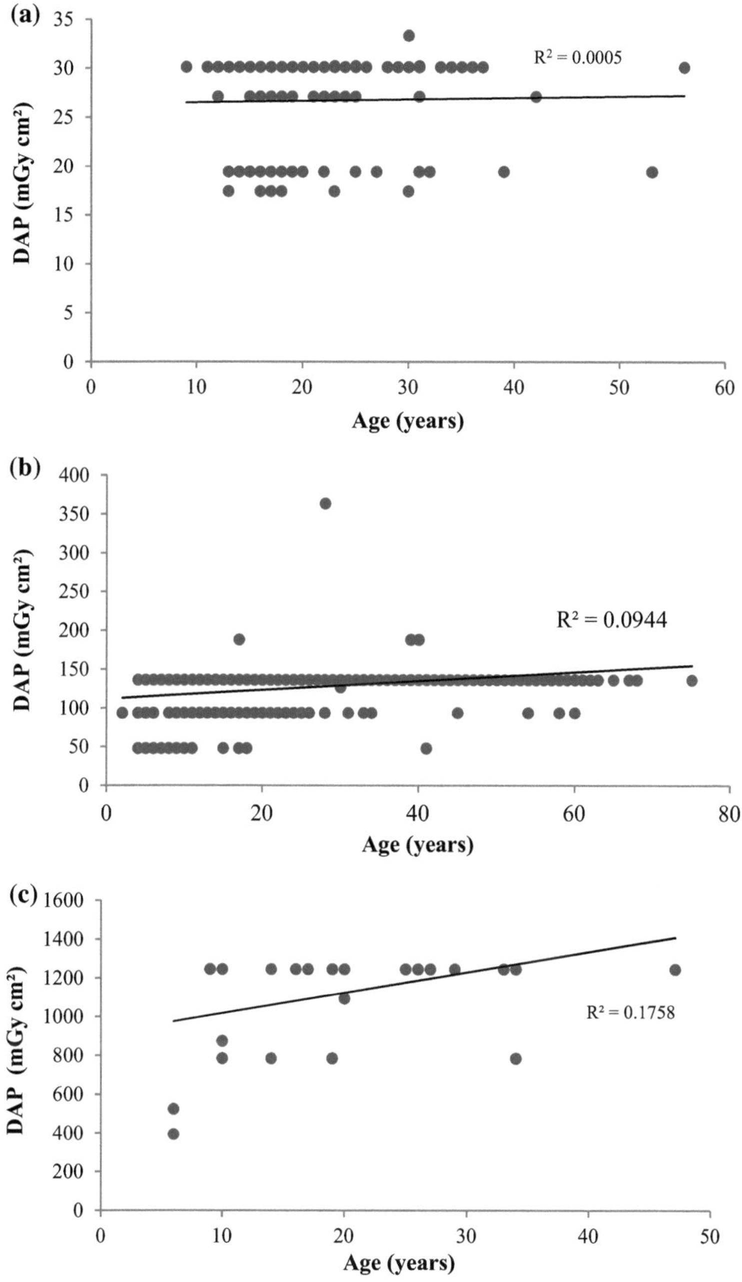

Fig. 2 Distribution of DAP based on patient's age for each dental radiography: **a** cephalometric examination **b** OPG examination **c** CBCT examination

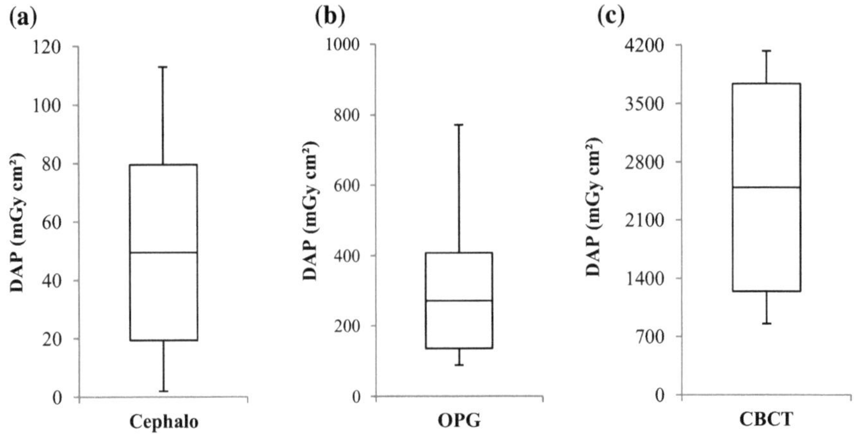

Fig. 3 Box and whisker chart of DAP values: **a** OPG **b** CBCT **c** cephalometric examination

represented by the correlation line. From the graph, it can be observed that the dose received by patient were increased gradually with age.

From the statistical analysis, most procedures showed a positive correlation with values of R^2 of 0.0944, 0.1758, 0.0005, for OPG, CBCT, and cephalometric examinations, respectively. The dose for CBCT examination showed the best line compared to others. However, the DAP values were dependent on the acquisition parameters such as FOV, kVp, and mAs.

The mean, maximum, minimum, first and third quartile for the collected DAP of each dental procedure were presented as box and whisker plot in Fig. 3. From Figs. 2 and 3, it shows that DAP values received from CBCT were the highest among other procedures with mean DAP value of 1245.4 mGy cm^2 of dose per radiograph, followed by OPG procedure with mean DAP of 136.0 mGy $cm^{2.}$ and mean DAP for cephalometric was 30.1 mGy cm^2. However, the calculated third quartile value for all procedures were similar with the mean value of collected DAP.

The DAP values for CBCT examination ranged from 392.4 to 1245.4 mGy cm^2, a range from 47.4 to 363.2 mGy cm^2 for OPG examination, and 17.4 to 33.3 mGy cm^2 for cephalometric examination.

3.2 *Comparative Study of Local Dose with the International DRLs*

For comparative study, the histogram of collected DAP against the frequency of cases were plotted to demonstrate the distribution of patient dose in comparison with the selected international DRL as the reference values (Fig. 4). The solid vertical line

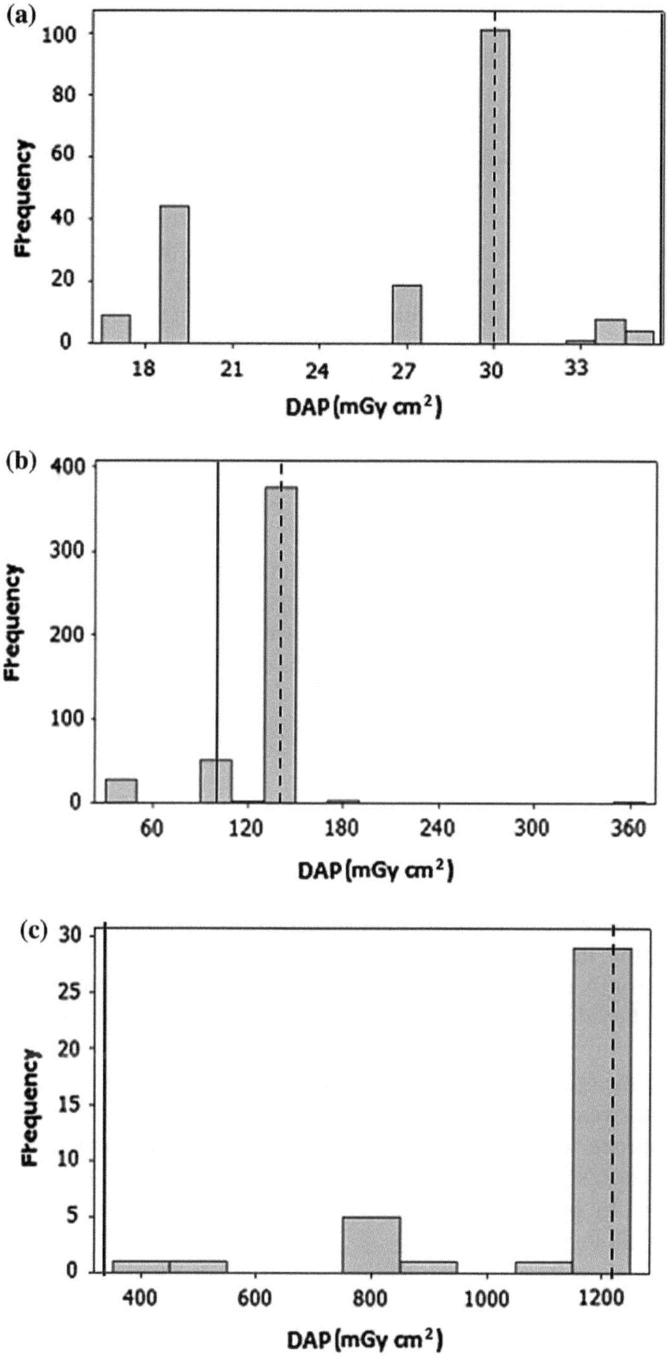

Fig. 4 The distribution of DAP received for each examination: **a** cephalometric **b** OPG and **c** CBCT in comparison with the established DRL

indicates the selected international DRLs [7–9] and the dotted vertical line indicates the third quartile of the collected local dose data.

Based on Fig. 4, it shows that the doses received from cephalometric examination were below the reference value. The calculated third quartile value of 30.1 mGy cm^2 for cephalometric were below the reference value (40 mGy cm^2). For OPG procedure, some dose data were below the selected DRL but most of the DAP and third quartile values were exceeded the suggested international DRLs.

For CBCT examination, all DAP received were higher than the suggested achievable dose. The third quartile value of DAP (124.5 mGy cm^2) was also higher than the suggested value. From the dose investigation, the exceeding values of local dose data for both OPG and CBCT might due to different exposure setting such as selection of tube voltage and current, exposure time, and FOV used during the examination. In this study, the tube voltage used were ranged between 64 to 84 kV for OPG.

For CBCT, the tube voltage used were fixed at 90 kV and the exposure time were ranged between 12 to 14 s. Thus, the exposure parameters must be optimized while ensuring adequate image quality as to minimize the dose received by patient [10, 11].

In this study, the doses were compared with the reference value for adult patient due to limited number of cases involving paediatric patients at AMDI USM, Malaysia. Few studies had proposed the reference dose values for dental radiography of paediatric patients [8, 9, 12] and can be used as the reference for comparative study.

4 Conclusion

This study shown that the dose received by patients undergo OPG examination depends on the exposure parameters (kV, mA and s). However, the DAP value for CBCT examination decreased when the mAs and FOV is reduced. For cephalometric examination, the dose depends on the kV and exposure time, but not on the tube current as the mA is fixed for each radiograph. Thus, it shows that the patient dose depends on the selection of image acquisition protocol during exposure.

The comparative study of the local dose data with international DRL shows that the dose for the OPG and CBCT examinations were exceeded the reference value, but the doses were still within the dose range for the cephalometric examination. Thus, a proper review on local clinical practice should be performed for corrective plan for the doses that exceed the range from reference value. The assessment of dose received by patient is important for justification and optimization of dose in diagnostic radiology practice.

Further work will include results involving paediatric patients and comparative study with the current established DRLs for paediatric patients in dental imaging.

Acknowledgements The authors would like to thank all radiographers and staff at Imaging Unit, Clinical Trial Complex, AMDI USM, Malaysia for their help and support throughout this work.

References

1. Kim EK, Han WJ, Choi JW, Jung YH, Yoon SJ, Lee JS (2012) Diagnostic reference levels in intraoral dental radiography in Korea. Imaging Sci Dent 42(4):237–242. https://doi.org/10.5624/isd.2012.42.4.237
2. Shin H, Nam K, Park H, Choi H, Kim H, Park C (2014) Effective doses from panoramic radiography and CBCT (cone beam CT) using dose area product (DAP) in dentistry. Dentomaxillofacial Radiology 43(5):20130439
3. IAEA (1996) International basic safety standards for protection against ionizing radiation and for the safety of radiation sources
4. ICRP (1996) ICRP Publication 73: radiological protection and safety in medicine, vol. 22. Elsevier Health Sciences
5. ICRP (2011) Cone beam CT for dental and maxillofacial radiology. Provisional guidelines A report prepared by the SEDENTEXCT project. Version 1. http://www.sedentexct.eu
6. ICRP (1999) Guidance on diagnostic reference levels (DRLs) for medical exposures: radiation protection 109: Directorate-General Environment, Nuclear Safety and Civil Protection, European Commission
7. Hart D, Hillier MC, Shrimpton PC (2012) Doses to patients from radiographic and fluoroscopic x-ray imaging procedures in the UK—2010 review. HPA-CRCE-034. http://www.hpa.org.uk/Publications/Radiation/CRCEScientificAndTechnicalReportSeries/HPACRCE034
8. Holroyd JR (2011) National reference doses for dental cephalometric radiography. Br J Radiol 84(1008):1121–1124
9. SEDENTEXCT Project (2011) Radiation protection: cone beam CT for dental and maxillofacial radiology. Evidence based guideline. www.sedentexct.eu/system/files/sedentexct_project_provisional_guidelines.pdf
10. Drage N, Walker A (n.d.). Faculty of general dental practice (UK). Radiation doses and risks in dental practice. http://www.fgdp.org.uk/publications/selection-criteria-for-dental-radiography1/2-use-of-ionising-radiation-/21-radiation-doses-and-risks-in-dental-practice.ashx
11. Kiljunen T, Kaasalainen T, Suomalainen A, Kortesniemi M (2015) Dental cone beam CT: a review. Phys Med 31(8):844–860
12. Kim Y, Yang B, Yoon S, Kang B, Lee J (2014) Diagnostic reference levels for panoramic and lateral cephalometric radiography of Korean children. Health Phys 107(2):111–116

Evaluation of Metal Artifacts from Stainless Steel and Titanium Alloy Orthopedic Screw in Computed Tomography Imaging

D. Yusob, J. Zukhi, A. A. Tajuddin and R. Zainon

Abstract Artifacts arising from metallic implant had been a concern for Computed Tomography (CT) imaging in obtaining optimal image quality. The main aim of this study was to evaluate the metal artifacts severity from two different types of orthopedics screw and to optimise CT imaging parameters for metallic implants. A water-based abdomen phantom of diameter 32 cm (adult body size) was fabricated using polymethyl methacrylate (PMMA) materials. The fabricated phantom was scanned with dual-energy CT at 80 and 140 kV, and single-energy CT at 120 kV. Two types of orthopedic screws; titanium alloy (grade 5) and stainless steel (grade 316L) was used in this study. A phantom with orthopedics metal screw was scanned at various pitch (0.35, 0.60, 1.20) and slice thickness of 1.0, 3.0, 5.0 mm. The tube current was applied automatically using tube current modulation. In this phantom study, the severity of stainless steel and titanium alloy was analysed. Results showed that the signal-to-noise ratio (SNR) of titanium alloy was higher than the SNR of stainless steel. The optimal image quality of metallic implant was obtained at imaging parameters of pitch at 0.60 and 5.0 mm slice thickness. The use of optimum CT imaging parameters for orthopedic screw resulted in an improved CT image, as the SNR increases. This finding proves that optimum CT imaging parameters are able to reduce the metal artifacts severity on CT images. Therefore, it has potential for improving diagnostic performance in patients with severe metallic artifacts.

1 Introduction

Computed Tomography (CT) is a modality which is widely used as medical imaging due to the combination of multiple angles of x-ray projections taken to produce detail

D. Yusob · J. Zukhi · R. Zainon (✉)
Oncological and Radiological Sciences Cluster, Advanced Medical and Dental Institute, Universiti Sains Malaysia, Bertam, 13200 Kepala Batas, Pulau Pinang, Malaysia
e-mail: rafidahzainon@usm.my

A. A. Tajuddin
School of Physics, Universiti Sains Malaysia, 11800 Minden, Pulau Pinang, Malaysia

R. Zainon (ed.), *3rd International Conference on Radiation Safety & Security in Healthcare Services*, Lecture Notes in Bioengineering,
https://doi.org/10.1007/978-981-10-7859-0_5

image views of all types of body tissues. The dual-energy CT as a relatively new approach has brought about several advances in clinical CT interpretation, which provides advance clinical CT interpretation by providing the ability to generate monochromatic high energy quanta, which inherently remove the beam hardening artifacts [1, 2] and show fewer artifacts without changing the average energy. Recent development in Dual-Energy CT (DECT) improves the specificity of diagnostic information by displaying the presence of a specific substance based on their specific absorption spectrum [3].

However, the metal artifacts that arise from metallic implant during CT imaging cause a severe streak and shadow artifact in images. These metallic implants superimpose the structures of interest and deteriorate the image quality obtained, which reduced the image detailed and cause a limited diagnostic use [4, 5]. This is because the metallic implants have high density and atomic number, and is strongly attenuated to the x-ray photon during the diagnostic imaging. These lead to the production of severe image artifacts, due to the lack of data in the projection data obtained in CT images [6]. Therefore, due to the occurrence of metal artifacts, the diagnosis on CT images remains challenging with many cases rendered uninterpretable, even with hard convolution kernels [7].

A numbers of simple methods can be employed to reduce the severity of the metal artifacts by altering the parameters of the CT imaging protocol. The selection of CT imaging parameters also plays an important role in improving the image quality. However, these techniques can increase the radiation dose towards patient. Hence, the ALARA (As Low As Reasonably Achievable) principles must be applied to produce highest image quality with lowest possible radiation dose. Thus, the optimisation of CT imaging parameters for metallic implant is important to avoid giving the unnecessary radiation dose to the patient. Therefore, the purpose of this study was to evaluate the metal artifacts severity from two different types of orthopedics screw and to optimise CT imaging parameters for metallic implants.

2 Materials and Methods

2.1 Phantom Fabrication and Setup

A water-based abdomen phantom of diameter 32 cm (adult body size) was designed and fabricated using polymethyl methacrylate (PMMA) materials, which consisted of 7 cm diameter small cylinder tube to mimic the internal organ with metallic implants. An abdomen phantom was filled with water. Small PMMA tube cylinders were filled with the artifact-producing metal materials (titanium alloy and stainless steel) and then were placed inside the abdominal phantom. A water based-phantom with metal implants was scanned using CT scan parameters as shown in Table 1.

Table 1 Summary of CT parameters used

	Single-energy CT	Dual-energy CT
Tube voltage (kV)	120	80/140
Tube current	Tube current modulation	
Pitch	0.35, 0.60, 1.20	
Slice thickness (mm)	1.0, 3.0, 5.0	

2.2 CT Parameters

All CT examinations were performed using single-source CT (CT scanner SOMATOM Definition AS, Siemens Healthcare, Forchheim, Germany), with 128 slices system. Single-energy CT acquisitions were performed with the following parameters: pitch 0.35, 0.60, 1.20; tube voltage 120 kV, tube current modulation (care4Dose setting). Dual-energy CT acquisitions were performed with the following parameters: pitch 0.35, 0.60, 1.20; a tube voltage pair of 80 and 140 kV.

Single-energy CT datasets at 120 kV and dual-energy CT datasets at 80 and 140 kV were reconstructed with a slice thickness of 1.0, 3.0, 5.0 mm using a fixed field of view (FOV) of 200 mm (image matrix 512 × 512). A sharp convolution kernel was chosen for image reconstructions of the dual-energy and single-energy acquisitions. After image reconstruction were applied in the workstation, the image were then transferred to a workstation (Syngovia Workplace; Siemens Healthcare) for the mixed up of dual-energy CT data sets.

2.3 Image Analysis

The signal-to-noise ratio (SNR) of the CT images for the metal artifacts severity was evaluated. For calculation of the SNR, the Hounsfield unit values of the ROI were measured using five circular regions of interest (ROIs) over an area of 50 mm^2. The location of ROIs was standardised in the image analysis as shown in the Fig. 1.

2.4 SNR Measurements

The CT images obtained were analysed and the SNR were calculated. An average from five ROIs and background were used to calculate the SNR. The SNRs were calculated using equation show in Eq. 1. It is calculated by dividing the mean CT number of A by the standard deviation of the background [8].

$$\mathrm{SNR} = \frac{s_A}{\sigma_0} \tag{1}$$

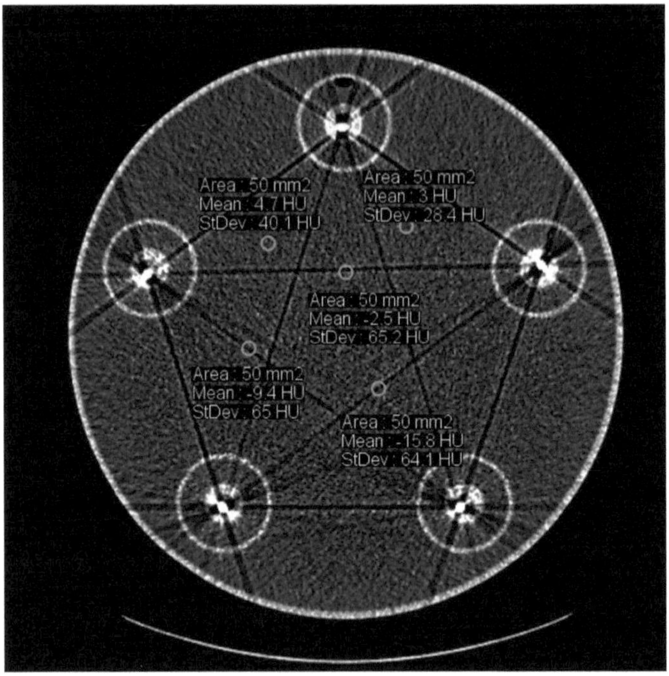

Fig. 1 The location of ROIs for CT image analysis

where, S_A was mean CT number of A in the ROI, and σ_0 (background noise) was the standard deviation of the background.

2.5 Statistical Analysis

The statistical analysis on the image quality was reported based on SNR measurement, and student's t-test was performed to determine significant differences between the results. A p-value < 0.05 was defined as statistically significant.

3 Results and Discussions

3.1 Single-Energy and Dual-Energy CT

Single-energy and dual-energy data set were evaluated and compared. Figure 2 shows SNR against slice thickness for single-energy and dual-energy CT. The SNR increases as the slice thickness increases for both data acquisitions. The SNR value obtained

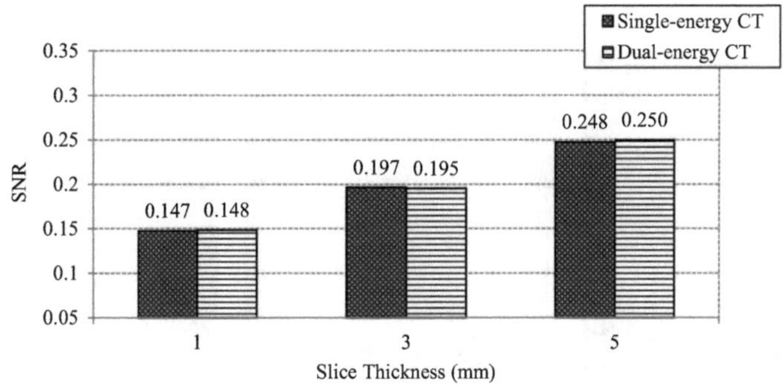

Fig. 2 The SNR of CT image obtained at different slice thickness but at the same pitch from single-energy (120 kV) and dual-energy CT (80/140 kV) scan modes

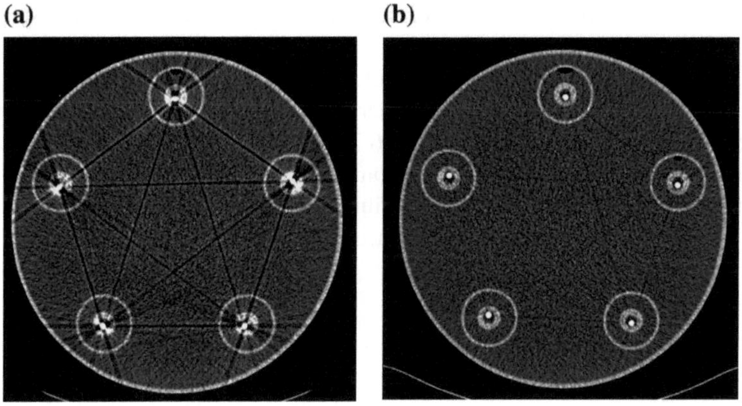

Fig. 3 The CT image of streaking artifacts produced from **a** stainless steel and **b** titanium alloy

at different slice thickness for both data acquisition is not significantly different with p-value of 0.5757. The same findings also stated by Exhibit et al. (2016) and Yusob et al. (2017), which stated that the CT image quality of the dual-energy CT is similar to the conventional single-energy obtained at 120 kV acquisitions [9, 10].

3.2 *Artifact Severity of Stainless Steel and Titanium Alloy*

Metal implant of stainless steel and titanium alloy produced different artifact severity in CT images. Figure 3 shows the CT images of different type of orthopedics screw materials. The CT images visualise the streaking artifacts vary with the presence of orthopedics screw materials. The streak artifacts are present between the orthopedics screw and elsewhere in the water phantom.

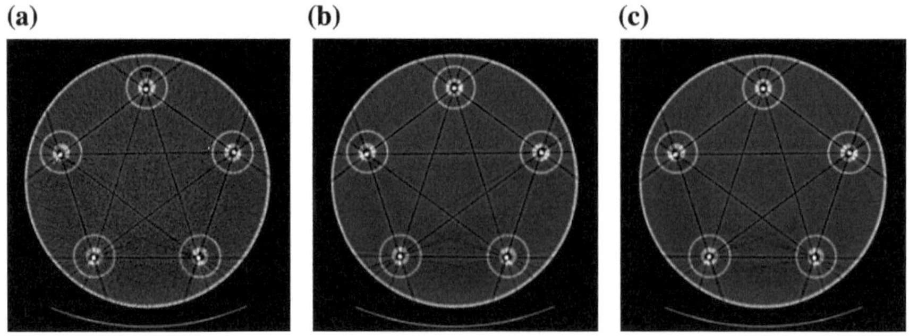

Fig. 4 CT images of stainless steel at slice thickness of **a** 1.0 mm, **b** 3.0 mm, and **c** 5.0 mm

Figure 3 shows the CT images of streaking artifacts from stainless steel and titanium alloy. Streaking artifacts are more pronounced with the presence of stainless steel screw than titanium alloy. The stainless steel produced more severe artifacts due to its higher atomic number and higher density than titanium alloy. The atomic number of stainless steel is 26.7 ($Z = 26.7$) while the atomic number of titanium alloy is 21.4 ($Z = 21.4$). Due to higher atomic number and density of stainless steel, it attenuates x-ray photon more than titanium alloy. Therefore, the streaking artifact on CT image produces more severe than titanium alloy.

3.3 *Effect of Slice Thickness on CT Images*

The slice thickness used in CT imaging parameters also affect the CT image. Figure 4 shows the CT image of stainless steel obtained at different slice thickness. The lowest amount of noisy image is obtained at slice thickness of 5.0 mm, whereas the highest amount of noisy image is at slice thickness of 1.0 mm. These noisy background decreases as the slice thickness increases. However, if the slice thickness if too thick, the image will be effected by partial volume effect (PVE). PVE is occurs when different objects are represented in the same voxel [11].

3.4 *The SNR at Different Slice Thickness and Pitch*

Each CT image obtained at different slice thickness and pitch were analysed and the SNR value was evaluated. The SNR is inversely proportional to the image noise. Therefore, the higher the SNR value, the better the CT image quality obtained. This is due to lower image noise produced. Figure 5 shows the SNR obtained at different pitch setting and slice thickness. The SNR increases as slice thickness increases and it decreases as the pitch increases.

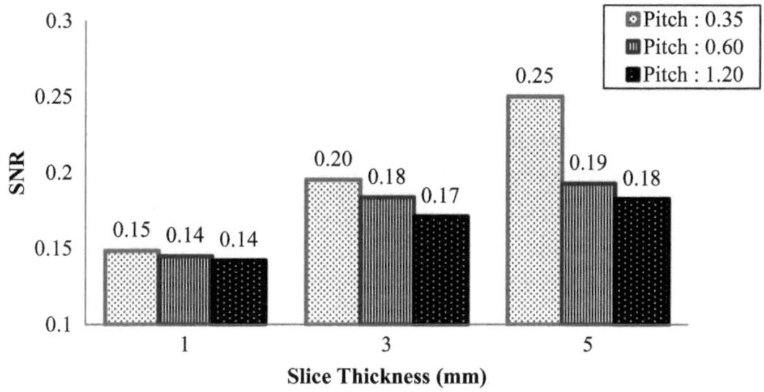

Fig. 5 SNR against slice thickness for pitch of 0.35, 0.60 and 1.20

Slice thickness of 5 mm has the highest SNR value compared to slice thickness of 1 and 3 mm, whereas, slice thickness of 1 mm shows the lowest SNR value as compared with the slice thickness of 3 and 5 mm. Pitch of 0.35 has the highest SNR value than pitch of 0.60 and 1.20. However, pitch of 1.20 shows the lowest SNR value as compared with the pitch of 0.35 and 0.60. The result shows the same trend for single-energy and dual-energy CT for both stainless steel and titanium alloy. The results obtained show the same trend for single-energy and dual-energy CT for both stainless steel and titanium alloy.

The CT image obtained from thinner slice thickness (1 mm) is noisier than thicker slice thickness (3 and 5 mm). This is due to the spatial resolution of thinner slice thickness is higher as compared to thicker slice thickness. Thinner the slice thickness gives a greater the spatial resolution of CT images. Thicker slice thickness gives an image with lesser spatial resolution. However, if the slice thickness is too thick, the images can be affected by artifact due to the PVE [12]. Therefore, a slice thickness of 5 mm is the optimum slice thickness parameters with lower noise, optimum spatial resolution and higher SNR value.

The SNR value decreases as pitch increases. Pitch affects both CT image quality and patients dose. Pitch is defined as the table distance in one 360° gantry rotation divided by beam collimation. Increasing pitch during scanning will increase the speed of moving patients through x-ray photon, thus will reduce patients dose. However, it also decreases the CT image quality obtained. This is due to the gaps in the x-ray photon where some tissue is not irradiated. While for the CT image with the lowest pitch will resulted in a good image quality. The radiation dose to the patients is higher due to the x-ray photon overlap and causes the same tissue being irradiated more than once per scan. Therefore, an optimum pitch of 0.60 is selected as it produced a good image quality (high SNR value) with lower radiation dose exposed to the patients.

4 Conclusion

In conclusion, we found that excessively thin slice thickness or high pitch settings resulted in degradation in diagnostic CT image quality or gives unnecessary radiation exposure. Although an increase in pitch settings leads to a lesser noise in CT images, however this will increase dose towards patient. Our results suggest that it is desirable to use is at pitch 0.60, and slice thickness of 5.0 mm. The image noise on CT images also can be reduced by altering the technical parameters of the CT protocol.

References

1. Bamberg F, Dierks A, Nikolaou K, Reiser MF, Becker CR, Johnson TRC (2011) Metal artifacts reduction by dual energy computed tomography using monoenergetic extrapolation. Eur Radiol 21:1424–1429
2. Meinel FG, Bischoff B, Zhang Q, Bamberg F, Reiser MF, Johnson TR (2012) Metal artifact reduction by dual-energy computed tomography using energetic extrapolation: a systematically optimized protocol. Invest Radiol 47:406–414
3. Johnson TR, Krauss B, Sedlmair M (2007) Material differentiation by dual energy CT: initial experience. Eur Radiol 17:1510–1517
4. Yu L, Li H, Mueller J, Kofler JM, Liu X, Primak AN, Fletcher JG, Guimaraes LS, Macedo T, McCollough CH (2009) Metal artifact reduction from reformatted projections for hip prostheses in multislice helical computed tomography: techniques and initial clinical results. Invest Radiol 44:691–696
5. Zhang X, Wang J, Xing L (2011) Metal artifact reduction in x-ray computed tomography (CT) by constrained optimization. Med Phys 38:701–711
6. Robert M, Michael G, David H (2004) Artifact analysis and reconstruction improvement in helical cardiac cone beam CT. IEEE Trans Med Imaging 23(9):1150–1164
7. Lee MJ, Kim S, Lee SA, Song HT, Huh YM, Kim DH, Han SH, Suh J (2007) Overcoming artifacts from metallic orthopedic implants at high-field-strength MR imaging and multi-detector CT. Radiographics 27:791–803
8. Murakami Y, Kakeda S, Kamada K, Ohnari N, Nishimura J, Ogawa M, Otsubo K, Morishita Y, Korogi Y (2010) Patient safety 64-section multidetector 3D CT angiography: evaluation with a vascular phantom with superimposed bone skull structures. AJNR Am J Neuroradiol 31(4):620–625
9. Exhibit E, Zhang C, Nicolaou S (2016) Dual-energy computed tomography (DECT): review of key principles and utility in common clinical settings Electronic Presentation Online Syst ECR2016/C-1624
10. Yusob D, Zukhi J, Tajuddin AA, Zainon R (2017) Evaluation of efficacy of metal artefact reduction technique using contrast media in computed tomography. J Phys Conf Ser 851:012009
11. Barrett JF, Keat N (2004) Artifacts in CT: recognition and avoidance. Radiographics 24:1679–1691
12. Primak AN, Mccollough CH, Michael R, Zhang RTRJ, Fletcher JG (2006) Relationship between noise, dose, and pitch in cardiac multi-detector row CT 1. Radiographics 26(6):1785–1795

Occupational Dose in Nuclear Medicine Department; Hospital Kuala Lumpur Experience

F. R. Kufian, M. K. A. Karim, N. N. Rapie, W. N. S. W. Aziz and S. Radziah

Abstract Personnel dose monitoring among nuclear medicine worker is the most vital component in the occupational safety. Generally, most of these workers are monitored by using variable type of personnel dosimeter such as optically stimulated luminescence (OSL), radio-photoluminescence (RPL), thermoluminescence dosimeter (TLD) and film badge. Therefore, we aimed to estimate whole body exposure of workers using OSL and RPL and evaluate the performance of both dosimeters in occupational dosimetry. 22 subjects (5 physicists, 6 pharmacists and 11 technologists) were participated in this study where both dosimeter was placed on mid-chest area for a period of 3 month. For analysis purpose, the RPL was sent to APM Nuclear Technology while OSL was analyzed by in house physicist using OSL reader (Microstar, Japan). Mean dose value Hp(10) of the nuclear medicine worker using OSL and RPL result obtained were 0.23 ± 0.11 mSv and 0.19 ± 0.07 mSv, respectively. As indicate from the result, there was a significant difference of dose between OSL and RPL. The range of dose value recorded for OSL and RPL measurement was 0.20 mSv to 0.26 mSv and 0.17 mSv to 0.21 mSv, respectively. The finding of this study shows that OSL was much more sensitive than RPL by a factor of 1.2. Therefore, OSL will enhance occupational safety program by minimizing radiation risk among radiation worker.

F. R. Kufian (✉) · N. N. Rapie · W. N. S. W. Aziz · S. Radziah
Department of Nuclear Medicine, Hospital Kuala Lumpur, Jalan Pahang, 50586 Kuala Lumpur, WP Kuala Lumpur, Malaysia
e-mail: razifskb@gmail.com

M. K. A. Karim
Department of Radiology, National Cancer Institute, Jalan P7, Precint 7 62250 Putrajaya, Malaysia
e-mail: khalis.karim@gmail.com

R. Zainon (ed.), *3rd International Conference on Radiation Safety & Security in Healthcare Services*, Lecture Notes in Bioengineering,
https://doi.org/10.1007/978-981-10-7859-0_6

1 Introduction

Nuclear medicine department in hospital utilize extensively the radiation from the radionuclide to diagnose and treat the human disease. It potentially poses an occupational health risk to the workers, as the personnel monitoring is the most important aspect need to undergo when working in a radiation environment. External exposure of personnel monitoring had been long introduced in Malaysia as enacted in ACT 304 personnel monitoring section paragraph 22(1) and (4) [1]. The primary objective of personnel monitoring is the assessment of occupational exposure delivered to personnel [2]. Personal dosimetry such optically stimulated luminescence (OSL), radio-photoluminescence (RPL), thermoluminescence dosimeter (TLD) and film badge is the tool to monitor occupational radiation exposures. Nowadays, the development and application of OSL is keep on growing and expanding. In Nuclear Medicine Department, Hospital Kuala Lumpur (NMHKL), film badge is the earliest dosimeter used since 1999. In 2011, TLD was introduced to replace the use of film badge. Then after a few years, RPL is used until mid of 2016 and currently NMHKL use OSL to monitor occupational exposure among radiation workers. As the technology evolves NMHKL seized the opportunity to perform a comparative dosimeter's field study based on different technologies namely RPL and OSL. Therefore, we aimed to estimate whole body exposures of workers using OSL and RPL and evaluate the performance of both dosimeters in occupational dosimetry.

1.1 *Optically Stimulated Luminescence*

The OSL made from a luminescent material which contain aluminium oxide (Al_2O_3:C). It possesses deep electron traps that get progressively filled by exposure to radiation. A fraction of the trapped electrons get transmitted by light exposure to luminescent site during the reading process. Typically 15% for doses of 1 mSv or less, the trapped electrons only needs a small proportion to be interrogated [3]. This condition will allow the dose to be read many times [3].

The fundamental behavior of the OSL is shown in Fig. 1. Aluminium Oxide (Al_2O_3) will emit blue light when it is stimulated by green light from the light source [4]. The amount of the exposure dose is directly proportional to the amount of blue light.

OSL reader (Micro star) used in this study which operate manually, it has dimension of 368 mm × 597 mm × 419 mm with approximate weight of 22.7 kg and equip with 36 LED array. It has capacity to read only single dosimeter at one time within 13 s plus handling time. OSL dosimeter use in this study is Inlight™ which designed for personnel monitoring of whole body using continuous wave OSL system [5]. The dosimeter consists of a plastic holder, which snaps shut to hold a plastic dosimeter packet. The dosimeter packet holds the metal or plastic filters and a plastic slide containing detector elements. The detector element is a layer of Al_2O_3 sandwiched

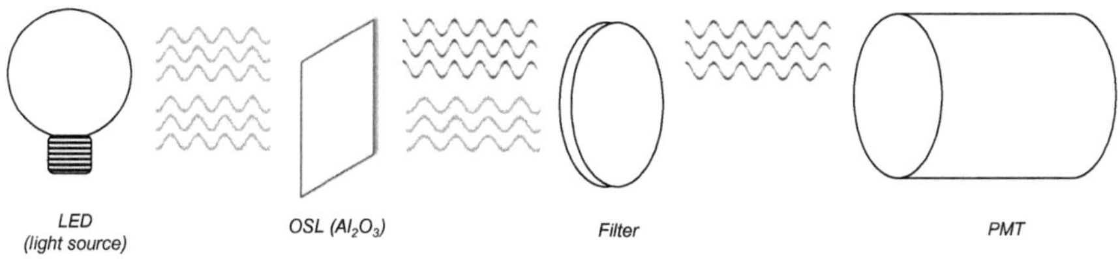

Fig. 1 Measurement principle of the OSL dosimeter

between two layers of polyester for a total thickness of 0.3 mm. It was designed for organization of institution that wishes to perform their own dosimetry [6].

1.2 Radio-Photoluminescence

RPL dosimeter is the photoluminescence technology based on silver activated phosphate glass produced by Chiyoda Technol Corporation. It also popularly known as 'glass dosimeters'. The RPL is analyzed by pulsed UV (~364 nm) light laser stimulation on glass element which will emit orange (~606 nm) luminescence and detected by RPL reader system [4]. RPL glass dosimeter model GD-450 is employed in this study. RPL made with compact design that has dimension of 45 mm × 13 mm × 5 mm and approximate weight of 5 g. Two types of plastic filters and three types of metal filters construction enable the measurement of gamma, X-ray and beta radiation with good precision.

2 Methodology

2.1 Study Design

This prospective dosimetry study is focused on measuring the mean personal equivalent dose Hp(10) received from daily duty of handling radioactive material. Generally, NMHKL offered diagnostic and therapeutic clinical service using high activity of Tc-99 m and I-131 in liquid form. In addition, few number of check sources which employ small activity of micro curie range such as Co-57, Ba-133 and Cs-137 in powder or solid form is used for equipment quality control purpose.

Table 1 Classification of working area for each profession

Group	Working area
Physicist	Hot laboratory, Gamma camera/SPECT CT room, radioactive waste room, decay tank, Radioiodine ward and physics laboratory
Pharmacist	Hot laboratory and QC room
Technologist	Hot laboratory, injection room and gamma camera/SPECT CT room

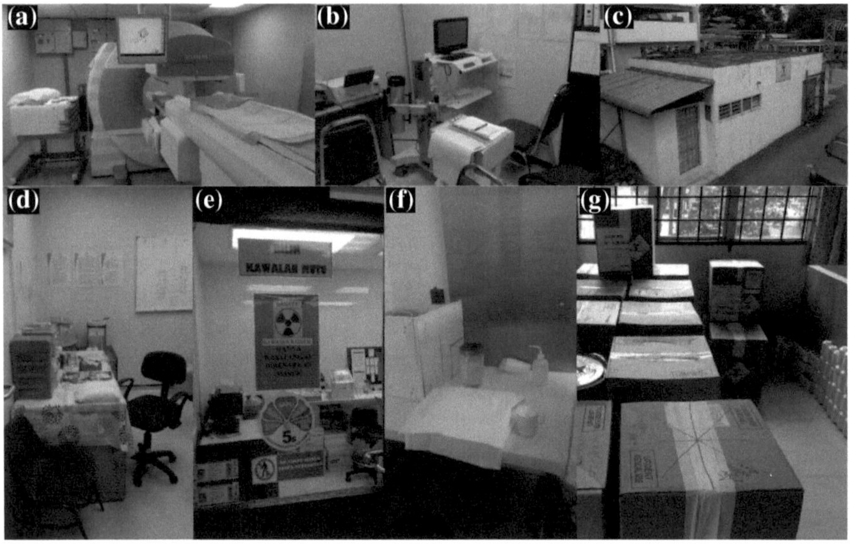

Fig. 2 Working area in NMHKL; **a** SPECT CT room, **b** physics laboratory, **c** decay tank, **d** injection room, **e** QC room, **f** Hot laboratory, **g** radioactive waste room, respectively

2.2 Subjects

Subjects involved in this study were five physicists, six pharmacists and eleven Technologists. All of them working actively in supervised and controlled area. Their working experience in NMHKL is varies from less than a year to more than 5 years (Table 1 and Fig. 2).

2.3 Dosimetry Procedure

This study was performed in a paired design with optically stimulated luminescence (OSL) dosimeter and radio-photoluminescence (RPL) dosimeter placed on mid chest area for a period of three months. The readings of OSL and RPL is anal-

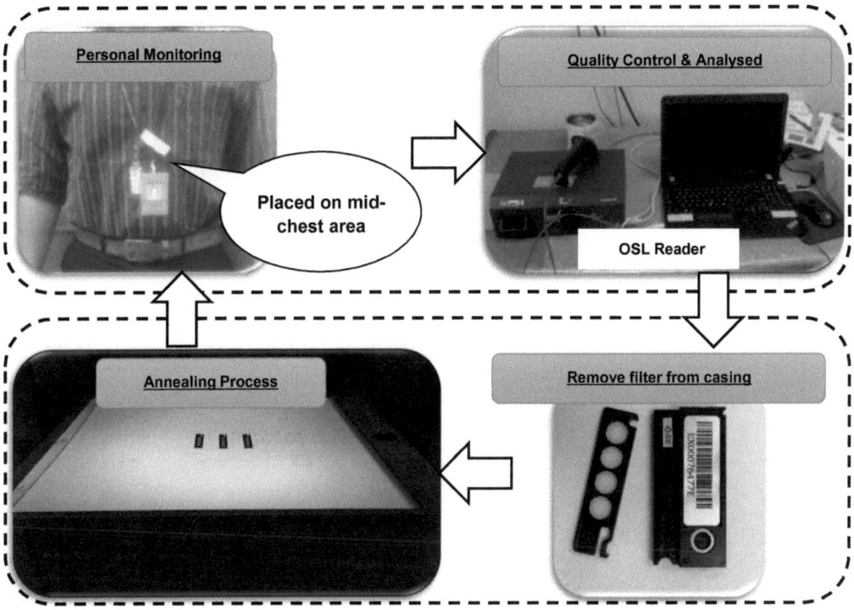

Fig. 3 Management of OSL

ysed by monthly basis. RPL (GD-450) were supplied and analyzed by Sinaran Utama Teknologi Sdn. Bhd. Malaysia. Meanwhile for XA Inlight™ OSL, we analyzed using OSL reader (Microstar) performed by in house physicist. Appropriate quality control of OSL reader is performed before analysis stage to conform the performance of the reader within stipulated standard or reference level establish during testing and commissioning stage. The annealing process of OSL is performed after complete the analysis to reset the dose reading of the dosimeter for multiple usage. OSL reader calibration is traceable to Glenwood Calibration Laboratory via calibration dosimeter irradiated to Cs-137 beam supplied by Laundauer Inc. Management of OSL has been summarized as shown in Fig. 3.

3 Result and Discussion

Data from the subject is normally distributed. The distribution of $Hp(10)_{OSL}$ and $Hp(10)_{RPL}$ for three month period are illustrated in Fig. 4, respectively for all subject. Arithmetic mean and standard deviation of the $Hp(10)_{OSL}$ and $Hp(10)_{RPL}$ are listed in the Table 2 respectively. The arithmetic mean of $Hp(10)_{OSL}$ is significantly higher ($p < 0.05$) than $Hp(10)_{RPL}$ by a factor of 1.2. Our finding shows that OSL response relatively to RPL is higher than 1.08 recorded in the previous study conducted in nuclear reactor environment [3]. Geometric means and coefficients of variance (CV)

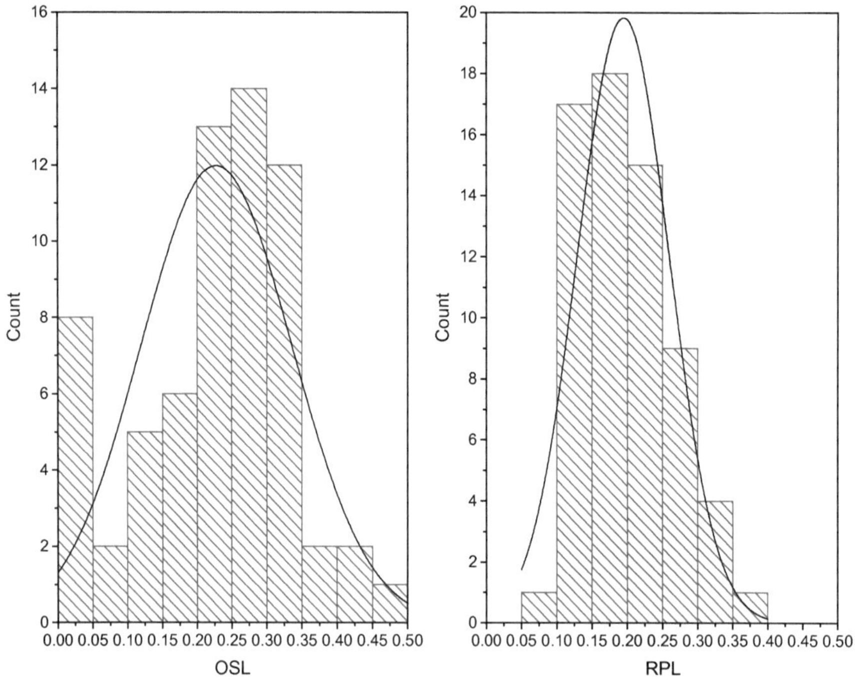

Fig. 4 Frequency distribution of $Hp(10)_{OSL}$ and $Hp(10)_{RPL}$ during 3 month period

Table 2 Distribution of radiation dose Hp(10) (mSv) (Whole)

	Radiation doses Hp(10) (mSv)	
	$Hp(10)_{OSL}$	$Hp(10)_{RPL}$
Arithmetic mean (SD)	0.23(0.11)	0.19(0.07)
Geometric mean (coefficient of variation, %)	0.18(47.8)	0.18(36.8)
Confidence level (95.0%)	0.03	0.02
Median (first; third quartiles)	0.24(0.17,0.3)	0.18(0.14,0.23)
Minimum; maximum	0.02;0.46	0.09;0.38

of $Hp(10)_{OSL}$ and $Hp(10)_{RPL}$ are shown in Table 2 respectively. Higher CV value of OSL compared to RPL indicate the less variation of the sample relative to their respective means. It may due to the higher detection limit around 50 μGy of OSL meanwhile for RPL is about 14 μGy [5]. This statement is supported with lower minimum value of $Hp(10)_{OSL}$ compared to $Hp(10)_{RPL}$ by a factor of 0.22 in Table 2. However, the post irradiation fading effect of RPL about 1% per month may contributed to lower arithmetic mean of RPL compare to OSL due to longer time needed before analysis performed in Sinaran Utama Teknologi Sdn. Bhd. Malaysia [5].

Table 3 Distribution of radiation dose Hp(10) (mSv) (Profession)

Profession	Radiation doses Hp(10) (mSv)	
	Mean $Hp(10)_{OSL}$ ± SD (Range)	Mean $Hp(10)_{RPL}$ ± SD (Range)
Physicist	0.16 ± 0.09(0.02–0.28)	0.14 ± 0.02(0.1–0.16)
Pharmacist	0.24 ± 0.08(0.04–0.34)	0.23 ± 0.06(0.14–0.32)
Technologist	0.24 ± 0.12(0.02–0.46)	0.20 ± 0.07(0.09–0.38)

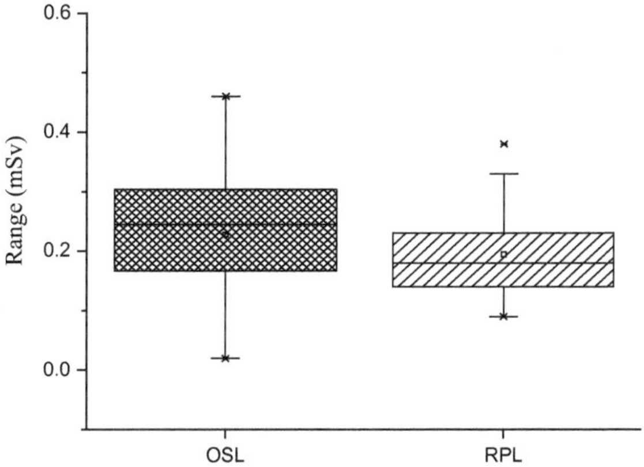

Fig. 5 Subjects distribution for $Hp(10)_{OSL}$ and $Hp(10)_{RPL}$

However, the mean and standard deviation of the $Hp(10)_{OSL}$ and $Hp(10)_{RPL}$ for each particular group; physicist, pharmacist and technologist are listed in Table 3 respectively. $Hp(10)_{OSL}$ and $Hp(10)_{RPL}$ during period of study for all group did not show any significance difference ($p > 0.05$). There are significant fair correlation between $Hp(10)_{OSL}$ and $Hp(10)_{RPL}$ for all subject during period of study. It may due to the high detection limit of OSL compared to RPL. Futher details of investigation on each particular group reveal mix outcome whereas only $Hp(10)_{OSL}$ and $Hp(10)_{RPL}$ of technologist show significant fair correlation while the remaining group did not show any significant weak correlation [7]. There are different outcome among various group under consideration due to their nature of work, frequency, duration working hour or occupancy factor at different working area.

The box plot of Hp(10) illustrated in Fig. 5 shown overall pattern of distribution for $Hp(10)_{OSL}$ and $Hp(10)_{RPL}$. The median, first quartiles and third quartiles of $Hp(10)_{OSL}$ and $Hp(10)_{RPL}$ are listed in Table 2 and Fig. 5. $Hp(10)_{OSL}$ is consistent with normal distribution while $Hp(10)_{RPL}$ data is unlikely normally distributed due to present of an outlier. There are one data point of outlier spotted in the box plot of $Hp(10)_{RPL}$. $Hp(10)_{OSL}$ of that particular point also recorded almost similar value. The detail investigation shown that the outlier data is contributed from junior staff with working experience of less than 5 years (Tables 4 and 5).

Table 4 Distribution of subject involved in the study according to their experience

No of year's experience	No of subject
Less than 1 year	5
Between 1 to 5 years	13
More than 5 years	4

Table 5 Statistical analysis for radiation dose Hp(10)

Statistical analysis	Whole	Group		
		Physicist	Pharmacist	Technologist
Mean T-Test (two tail)	$p < 0.05$	$p > 0.05$	$p > 0.05$	$p > 0.05$
Pearson correlation	0.493 ($p < 0.05$)	0.311 ($p > 0.05$)	0.222 ($p > 0.05$)	0.517 ($p < 0.05$)

4 Conclusion

This study represents an actual in field intercomparison dosimetry performance of RPL and OSL at the medical facility set up particularly NMHKL. The arithmetic mean of the $Hp(10)_{OSL}$ is significantly higher than $Hp(10)_{RPL.}$ Our finding is agreed with the previous study done by other researcher. Higher sensitivity of the OSL offer better estimation for occupational exposure. Therefore, OSL will enhance occupational safety program by minimizing radiation risk among radiation worker.

Acknowledgements Special thanks to our beloved Head of Department, Dr. Siti Zarina binti Amir Hassan for the encouragement to conduct this study. We also would like to thanks Nuclear Medicine Department, Hospital Kuala Lumpur for staffs cooperation and facilities support.

References

1. Law of Malaysia, Act 304: Atomic Energy Licensing Act 1984
2. Protection R (1991) ICRP publication 60. Ann ICRP, 21(1.3)
3. Garcier Y, Cordier G, Pauron C, Fazileabasse J (2006) Intercomparison of passive dosimetry technology at EDF facilities in France. Radiat Prot Dosimetry 124(2):107–114
4. Perks CA, Yahnke C, Million M (2008) Medical dosimetry using optically stimulated luminescence dots and microStar readers. In: 12th international congress of the international radiation protection association
5. Bhatt BC (2011) Thermoluminescence, optically stimulated luminescence and radiophotoluminescence dosimetry: an overall perspective. Radiat Prot Environ 34(1):6
6. Bøtter-Jensen L, McKeever SW, Wintle AG (2003) Optically stimulated luminescence dosimetry. Elsevier
7. Al-Abdulsalam A, Brindhaban A (2014) Occupational radiation exposure among the staff of departments of nuclear medicine and diagnostic radiology in Kuwait. Med Princ Pract 23(2):129–133

Determination of Mass Attenuation Coefficient of Paraffin Wax and Sodium Chloride as Tissue Equivalent Materials

N. A. Baharul Amin, N. A. Kabir and R. Zainon

Abstract The main goal of this study was to determine the mass attenuation coefficient of paraffin wax and sodium chloride as tissue equivalent materials in medical imaging. In this study, a series of attenuation coefficient was evaluated for a mixture of paraffin wax and NaCl to be established as tissue equivalent materials. Five samples with difference paraffin-to-NaCl ratio were prepared. The attenuation coefficient of each sample was evaluated within energy range of 0.662 and 1.33 MeV. The measurements were performed with a NaI(Tl) detector for gamma spectrometry. The attenuation coefficient values obtained from this experiment were compared with mass attenuation coefficient obtained from the XCOM software and ICRU Report 44. The mass attenuation coefficients of the selected sample and human soft tissue are 0.151 cm^2/g and 0.149 cm^2/g respectively at energy of 150 keV. A sample with 25% of NaCl used was selected as an ideal tissue-equivalent material in this study.

1 Introduction

Radioactive has been used in both medicine and industry since more than a century ago [1]. Understanding the interaction of radiation is crucial as the use of radioactive especially in medicine field grows rapidly. One of preceding application to study the radiation interaction with human body is the implementation of tissue equivalent material. Studies on tissue equivalent materials for are necessary for radiation dosimetry, calibration and medical imaging [2].

Many studies on difference chemicals used to represented tissue-equivalent materials were constructed [3, 4] to ensure the level of similarities interaction of ionizing radiation to human tissue. In order to be accepted as a tissue-equivalent material,

N. A. Baharul Amin · R. Zainon (✉)
Oncological and Radiological Sciences Cluster, Advanced Medical and Dental Institute, Universiti Sains Malaysia, Bertam, 13200 Kepala Batas, Pulau Pinang, Malaysia
e-mail: rafidahzainon@usm.my

N. A. Baharul Amin · N. A. Kabir
School of Physics, Universiti Sains Malaysia, 11800 Minden, Pulau Pinang, Malaysia

R. Zainon (ed.), *3rd International Conference on Radiation Safety & Security in Healthcare Services*, Lecture Notes in Bioengineering,
https://doi.org/10.1007/978-981-10-7859-0_7

the absorbed and the scattered radiations should be as close as possible to those experienced by the irradiated tissue under similar conditions [5]. To understand the physical properties of tissue equivalent materials, a study of mass attenuation coefficient parameters is important.

Mass attenuation coefficient is a reasonable way for evaluating the usage of tissue-equivalent materials [6]. Mass attenuation coefficient is defined as the measure of a material's ability to absorb or scatter electromagnetic radiation in any form (e.g., x-rays, gamma ray, etc.) per unit of mass. The total mass attenuation coefficient can be calculated through the photon energy and constituent elements of the material [7, 8]. The mass attenuation coefficient is then can be used to derive other photon interaction parameters. Linear attenuation coefficient describes the fraction of a beam of x-rays or γ-rays that is absorbed or scattered per unit thickness of the absorber. In order to measure gamma ray attenuation coefficient, Beer–Lambert Law was used by employing standard transmission method with narrow beam geometry [9].

The linear attenuation coefficients were determined by using the Beer-Lambert's formula [9]:

$$I = I_o e^{-\mu x} \tag{1}$$

where,

μ = linear attenuation coefficient (cm^{-1})
I_o = initial intensity
I = incident intensity
x = thickness of the sample (cm)

The maximum errors in these coefficients were calculated from errors initial intensities, I_o, incident intensities, I and densities, ρ using the following equation;

$$\Delta\left(\frac{\mu}{\rho}\right) = \frac{1}{\rho x}\left[\frac{\Delta I_o}{I_o} + \frac{\Delta I}{I} + \frac{\Delta\rho}{\rho}\ln\left(\frac{I_o}{I}\right)\right] \tag{2}$$

Where,

x = Thickness of sample (cm)
ΔI_0 = errors in the intensity I_0
ΔI = errors in the intensity I
$\Delta\rho$ = errors in the density, ρ, respectively.

2 Experimental Design

Five samples were prepared by melting solid paraffin at 60 °C on a hotplate. Then, NaCl was added into the melted paraffin wax. The mixture was then placed into the $3 \times 3 \times 0.5$ cm aluminium mould. Different ratio paraffin-to-NaCl was prepared to find the most suitable amount of NaCl added into paraffin to simulate the human soft tissue. The paraffin-to-NaCl ratio was prepared as shown in Table 1:

Table 1 The ratio of paraffin-to-NaCl in percentage for each sample

Sample	Paraffin wax (%)	NaCl (%)
A	100	0
B	90	10
C	80	20
D	75	25
E	70	30

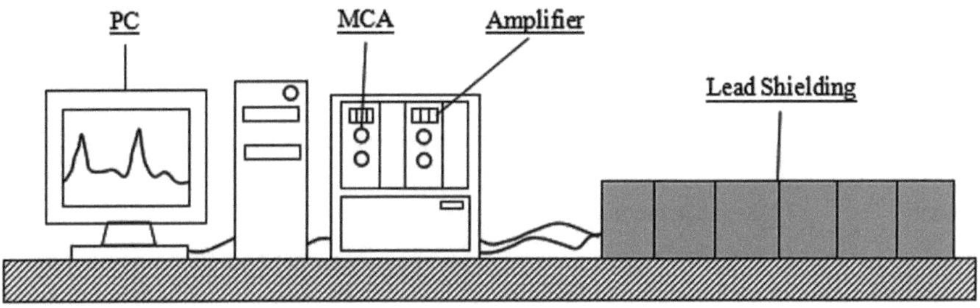

Fig. 1 A schematic diagram of experimental setup shown the amplifier, MCA and PC. The scintillation system used was placed in the lead shielding structure

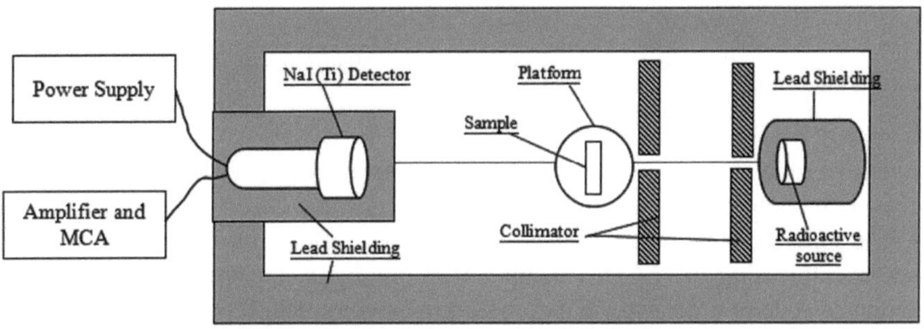

Fig. 2 A schematic diagram of scintillation system in the lead shielding structure

The scintillation system used in this study consists of 1.5" × 1.5" NaI(Tl) detector, amplifier, multichannel analyser (MCA) and power supply as shown in Figs. 1 and 2. The NaI(Tl) detector and source were both shielded with adequate lead housing to reduce the scattered radiation coming directly from the source or from the background. Five radioactive sources were used including Cs^{137}, Mn^{54}, Na^{22} and Co^{60}. The energies range used in this study was between 0.662 and 1.33 MeV. These energies were chosen to represent the energy range used in radiotherapy and nuclear medicine.

The samples were placed between the source and the detector. The distance between sample and detector was 20 cm and distance between sample and source was 10 cm. The real time was set for 6000 s. The photo peak, full width at half maximum (FWHM) and net area of photo peak were measured using Meastro software. The

Table 2 Mass attenuation coefficient of each sample prepared in this study for energy range of 662–1173 keV

Sample	Mass attenuation coefficient (μ/ρ) in cm^2/g					
	662 keV		834.85 keV		1173.24 keV	
	Expt.	XCOM	Expt.	XCOM	Expt.	XCOM
A	0.08710	0.08600	0.07730	0.07900	0.06930	0.06600
B	0.08560	0.08600	0.07600	0.07800	0.06810	0.06500
C	0.08410	0.08500	0.07470	0.07600	0.06690	0.06450
D	0.08260	0.08600	0.07400	0.07500	0.06640	0.06300
E	0.08430	0.08500	0.07340	0.07500	0.06580	0.06400

Table 3 Mass attenuation coefficient of each sample prepared in this study for energy range of 1274–1332 keV

Sample	Mass attenuation coefficient (μ/ρ) in cm^2/g			
	1274.53 keV		1332.50 keV	
	Expt.	XCOM	Expt.	XCOM
A	0.06360	0.06320	0.06060	0.06100
B	0.06250	0.06200	0.05950	0.06000
C	0.06150	0.06160	0.05850	0.05950
D	0.06640	0.06300	0.05790	0.05900
E	0.06040	0.06000	0.05740	0.05800

density of each sample was measured by Archimedes principle using distilled water as the immersion fluid.

The mass attenuation coefficient values were compared with those calculated theoretically using the XCOM computer program and database [9]. The XCOM computer code is a database which uses pre-existing data bases for coherent and incoherent scattering, photoelectric absorption, and pair production cross sections to calculate mass attenuation coefficients at photon energies of 1 keV–1 GeV (Berger and Hubbell 1987).

3 Result and Discussion

From the precisely measured densities of these tissue-equivalent materials samples, the mass attenuation coefficients were then obtained and results are shown in Tables 2 and 3. The mass attenuation coefficients for each samples were calculated and the results were compared with the mass attenuation coefficient from XCOM. Results showed that there is a good agreement between both measurement and theoretical values.

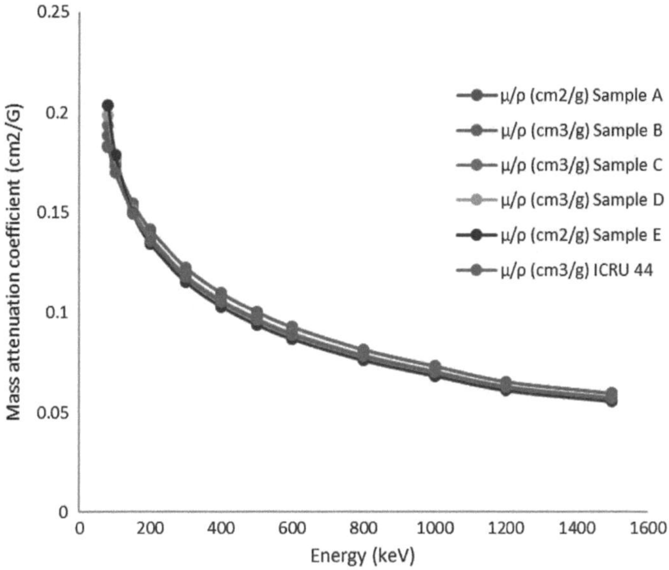

Fig. 3 Mass attenuation coeffcient of each sample prepared in this study at different energy range

Figure 3 shows that the mass attenuation coefficient increases with the increment of NaCl ratio in the mixture. The slopes of the curves are different at different energies due to the dominance of partial interaction processes in energy regions under consideration. Also, corresponding to a fixed composition, mass attenuation coefficients decreases as the gamma-ray energy increases. The energy dependence of photon interaction with the sample can be seen on Fig. 3. This is due to different photon absorption mechanism occurs at different energy range. Photoelectric effect occurs at low energy, Compton scattering at low and mid-energy range and pair production interaction at high energy (>1022 keV energy).

The mass attenuation coefficient of samples study was compared with mass attenuation coefficient of human soft tissue based on ICRU Report 44. A good agreement achieved between the mass attenuation coefficient of samples and ICRU Report 44. Sample D was chosen in this study as the most suitable mixture for tissue equivalent materials.

4 Conclusion

In conclusion, the NaCl increases the mass attenuation coefficient of paraffin wax. A suitable ratio of paraffin-to-NaCl and paraffin wax can be used as tissue-equivalent material in the energy range of 0.662 and 1.33 MeV. Sample D with 25% of NaCl used was selected as an ideal tissue-equivalent material in this study.

References

1. Reed AB, Killewich LA (2011) J Vasc Surg 53(1):3S–5S
2. Naderi SM, Sina S, Karimipoorfard M, Lotfalizadeh F, Entezarmahdi M, Moradi H, Faghihi R (2015) Radiat Prot Dosim:1–6
3. Chitralekha A, Kerur BR, Lagare MT, Nathuram R, Sharma DN (2005) Radiat Phys Chem 72:1e5
4. Baltas H, Celik S, Cevik U, Yanmaz E (2007) Radiat Meas 42:55e60
5. Rao ASM, Abebe G (2015) Int J Adv Res Sci Eng 4(1):2319–8354
6. Jones AK, Hintenlang DE, Bolch WE (2003) Med Phys 30:2072–2081
7. Hubbell JH (2006) Phys Med Biol 51:245–262
8. Sidhu BS, Dhaliwal AS, Mann KS, Kahlon KS (2012) Ann Nucl Energy 42:153–157
9. Berger MJ, Hubbell JH (1987) National Institute of Standards, Gaithersburg 1987NBSIR 87–3597, MD 20899 USA. http://physics.nist.gov/PhysRefData/Xcom/Text/XCOM.html

Study on Effectiveness of Physical Protection and Security Management of Radioactive Sources in Medical Institutions in Malaysia

M. N. M. Kamari, M. S. Yasir, Z. Kayun and P. Muthuvelu

Abstract Ministry of Health has joined the program of the Regional Security of Radioactive Sources (RSRS) to improve the Physical Protection and Security Management (PPSM) system of radioactive sources in medical institutions in Malaysia. This study is aimed to identify the compatibility of PPSM system components, used in medical institutions in Malaysia based on documents from the IAEA as well as to evaluate the effectiveness of its execution. Universiti Kebangsaan Malaysia Medical Centre, Ampang Hospital, Universiti Sains Malaysia Hospital, Advanced Medical and Dental Institute and Gleneagles Intan Medical Centre was chosen as a location for research. The system was assessed through questionnaires, interviews, observations and document review. From the results, it shows the effectiveness of implementing PPSM system still at a moderate level in all medical institutions in Malaysia. Improvements can be made through collaboration with the ministry and the management of medical institutions in which several components in the PPSM system has been identified as the details of the source of radioactive material, the security plan and security functions as well as to provide exposure to the employees by continuously organizing extensive courses, seminars and workshops.

1 Introduction

The September 11 attack at World Trade Center (WTC) in New York City in 2001 have caused 3,000 victims and killed all the 19 terrorists [1]. This event has been the starting point for the new millennium militant threats and early preventative measures need to be done to reduce its impact on the community in a country.

M. N. M. Kamari (✉) · Z. Kayun · P. Muthuvelu
Medical Radiation Surveillance Division, Ministry of Health, 62590 Putrajaya, Malaysia
e-mail: nathir.mkamari@moh.gov.my

M. N. M. Kamari · M. S. Yasir
Faculty of Science and Technology, School of Applied Physics, National University of Malaysia, 43600 Bangi, Selangor, Malaysia

R. Zainon (ed.), *3rd International Conference on Radiation Safety & Security in Healthcare Services*, Lecture Notes in Bioengineering,
https://doi.org/10.1007/978-981-10-7859-0_8

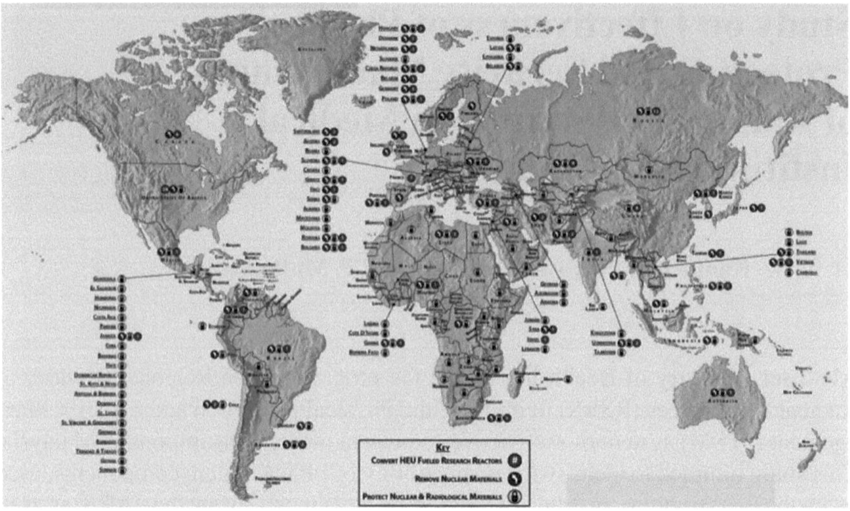

Fig. 1 Global Threat Reduction Initiatives (GTRI) around the world

1.1 Global Threat Reduction Initiatives-GTRI

The idea to develop Global Threat Reduction Initiatives (GTRI) which is introduced by the United States through US Secretary of Energy, Spencer Abraham on May 26, 2004 in Vienna is seen as a new global effort to address the issue of nuclear securities and threats of nuclear terror worldwide [2]. Figure 1 shows countries around the world that involved in Global Threat Reduction Initiatives (GTRI) and its activities.

The IAEA has outlined three key areas for international cooperation under Global Threat Reduction Initiative (GTRI) which is prevention, detection and response [3]. The IAEA will assist the affected countries in preventing the use of nuclear material or radioactive sources illegally by providing effective physical protection against nuclear materials during process of storage, transportation and protection at related nuclear facilities.

The IAEA also ensures that each country will has a system that can detect the beginning of any illegal activities related to nuclear materials or radioactive sources by giving education and training to customs and police officers, installing detection equipment at the national border and establishing a communications network with local and international law enforcement agencies to have a more effective information sharing [4].

Established an extensive collaboration between IAEA and government agencies and also other international organizations to ensure thorough and prompt actions for any illegal activities such as sabotages and terrorist acts involving nuclear or radioactive materials [5].

1.2 Regional Security of Radioactive Sources (RSRS)

In 2004, the Australian Government through the Australian Nuclear Science and Technology Organisation (ANSTO) has collaborated with countries from Southeast Asia to implement Regional Security of Radioactive Sources (RSRS) to combat the threat of terrorism and enhance technical cooperation between government agencies to develop and implement Physical Protection and Security Management (PPSM) system at facilities which has radioactive sources [6].

In 2009, the Ministry of Health Malaysia has participated in a Regional Security of Radioactive Sources (RSRS) under the Global Threat Reduction Initiatives (GTRI) with the collaboration of Australian Nuclear Science and Technology Organization (ANSTO) and National Nuclear Security Administrations (NNSA) to improve the Physical Protection and Security Management (PPSM) system for radioactive sources in medical institutions in Malaysia [7].

Among the activities conducted under this project are improvising the regulatory infrastructure to control the radioactive roots, provide training on securities management and physical protection of radioactive sources, conduct technical training programs and develop security plans and security procedures. All of these activities are implemented by adopting the IAEA Code of Conduct on the Safety and Security of Radioactive Sources [7].

Radioactive sources that involved are including Caesium-137 (Cs-137) for the use in blood irradiator and also Cobalt-60 (Co-60) and Iridium-192 (Ir-192) for gamma knife therapy and brachytherapy treatment machine.

1.3 Nuclear Security Summit (NSS)

In 2010, Nuclear Security Summit (NSS) was held to foster a global cooperation to fight against nuclear terrorism and control nuclear weapons from falling into the hands of terrorists. The discussion is mainly focused on identifying best approaches to combat threat of nuclear terrorism, protection for nuclear materials and related facilities and prevention of illicit trafficking of nuclear materials.

During the initial conference, Malaysia Prime Minister has expressed Malaysia's commitment on nuclear securities through the creation of the Strategic Trade Act 2010 (Act 708) which take effect on 1st July 2011. This Act allows Malaysia to be more effective and comprehensive in supervising and controlling exports, temporary loads, transit and brokerage of all strategic goods including weapons and related materials together with other activities that will or may facilitate the design, development and manufacture of nuclear and mass destruction weapons and its transmission system [8].

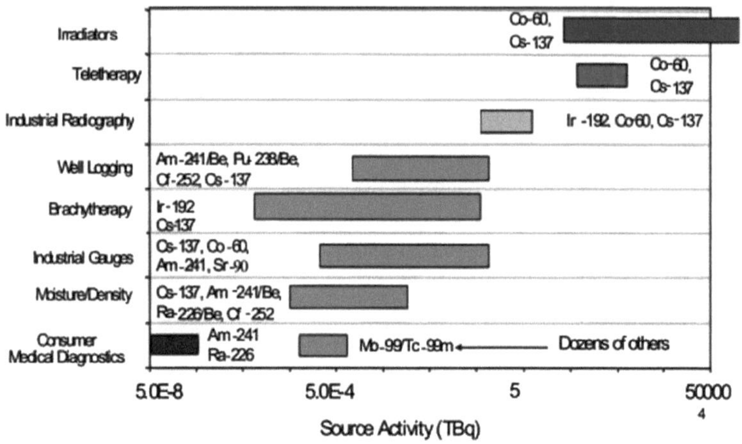

Fig. 2 Sample of fields which used sealed radioactive sources

1.4 Radioactive Source and Its Applications

Radioactive source is a radioactive substance that seals in a capsule or enclosed containers, in a solid form and non-exempted from regulatory control. It also covers a released radioactive if the source is leaked or broken. However, it will not covers encapsulated material for disposal purposes or nuclear material within the nuclear fuel cycle and energy reactor research area [9].

There are various sector that utilized sealed radioactive sources (Fig. 2) such as blood irradiator which using Caesium-137 (Cs-137) and brachytherapy treatment machine which contains Cobalt-60 (Co-60).

1.5 Categorization of Radioactive Source

Radioactive sources must be categorized since it has range of effects and a variety of potential hazards. Categorization of radioactive sources is used to determine the level of security of a radioactive source.

Through acute deterministic effects, the effects of radiation on health are worse for higher doses (Fig. 3) and has a potential of fatal or permanent injuries.

Categorization of radioactive sources is determined by considering ratio activities, A/D by using the formula below [10]:

$$\text{Aggregate } A/D = \sum_n \frac{\sum_i A_{i,n}}{D_n}$$

$A_{i,n}$ = activity of each individual source i of radionuclide n;
$D_n = D$ value for radionuclide n.

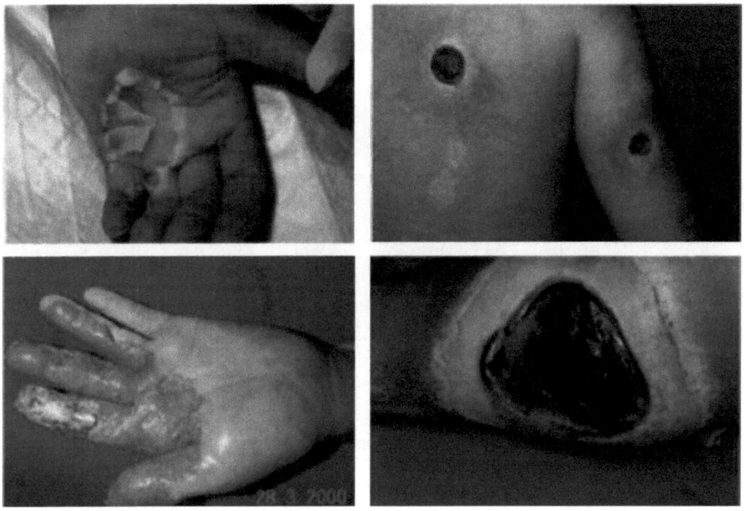

Fig. 3 Examples of acute deterministic effects

Category	Relative Danger	Activity Ratio (A/D)
1	Extremely Dangerous	A/D ≥ 1000
2	Very Dangerous	1000 >A/D ≥ 10
3	Dangerous	10 > A/D ≥ 1
4	Unlikely to be Dangerous	1 > A/D ≥ 0.01
5	Most Unlikely to be Dangerous	0.01 > A/D

Fig. 4 Relative danger for activity ratio, *A/D* by category

There are few factors need to be considered in categorizing radioactive source such as physical and chemical form of the radioactive source, analysis source, condition of the radioactive source and history of the accident. In Fig. 4 below are list of activity ratio categories, A/D and its relative danger.

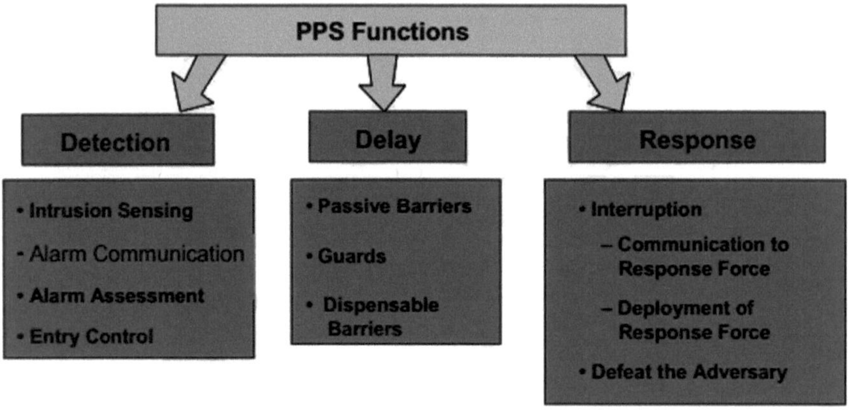

Fig. 5 Physical protection systems (PPS) function

1.6 Physical Protection System

Physical protection systems (PPS) is an integration system between people, procedures and equipment with the purpose of protecting the assets and related facilities from theft, sabotage or other malicious acts. It is designed to implement basic security functions such as deterrence, detection, delay, response and security management [11].

Figure 5 shows the security measures for the security function of the physical protection system. In order to develop a physical protection system, things like type of materials and assets to be protected, type of threat and the level of protection to be installed for the protection of the assets or materials need to be further analysed.

For a certain security concept, IAEA has proposed graded concept where the security level will be depending on the nature of the protected radioactive source. Graded concept are based on anticipation towards threats, relative attractiveness, effects from malicious actions and benefits of using radioactive sources.

1.7 Security Management

Security management will ensure security resources such as number of staff, equipment and amount of funding are adequate. It's also addressed few other matters including develop a policies, procedures, records, and plans for security resources and encourage for an active and effective securities culture [11].

There are several items consist in security management such as legislation compliancy for securities plans, threat assessment, basic threat design (DBT), graded concept approach, risk management, securities culture, confidentiality, ease of operation and nuclear securities as well as contingency plan [12].

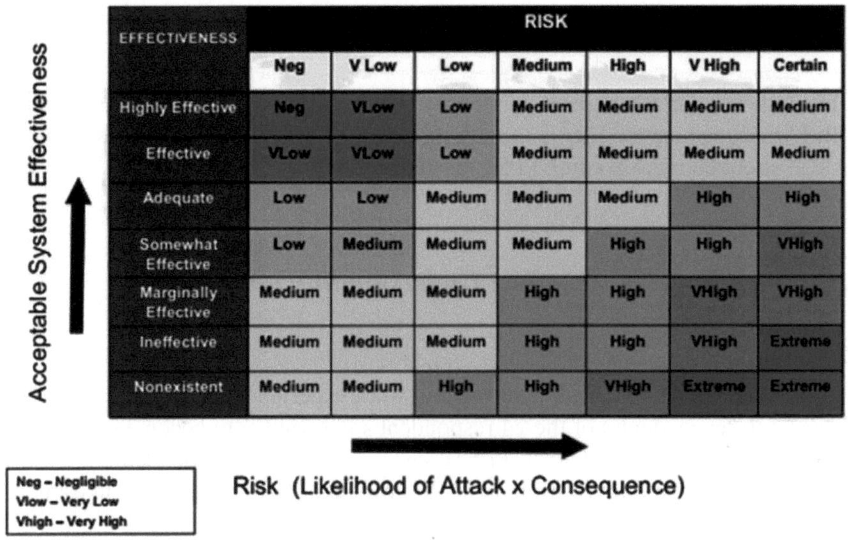

Fig. 6 The degree of the acceptable physical protection system effectiveness towards the level of risk

According to Khripunov [13], assessment on the security system of radioactive source can be carried out through 4 methods which is survey forms, interviews, observations and document reviews.

1.8 *Risk Management*

Zakariya and Kahn [14] has proposed the basic principles of risk management for the design of physical protection should consist of defense in depth, cultural security, quality assurance, confidentiality and contingencies.

Risk is measured based on probability, likelihood and consequences aspect [15]. Risk management involves things such as identifying, analyzing, evaluating, treating and monitoring risks. It is important to ensure that risks are always within a manageable range and identifying any resources that can be used for that purpose [16]. Figure 6 indicates the degree of the acceptable physical protection system effectiveness towards the level of risk [17].

2 Results and Discussions

Table 1 shows the comparison of security plan components in medical institution with IAEA documents. This is to identify potential enhancement plan that can be done to strengthen the existing PPSM system.

2.1 Work Experience Affect Understanding on the PPSM Procedure

Based on the results obtained, it shows that work experience does assist to increase 66.66% (16 respondents) of the 24 respondent's knowledge to become experts and capable to perform procedure in PPSM system. All the 16 respondents are those who have works experience from 11 to 20 years and 21 to 30 years as in Table 2.

Burke and Steensma [23] found that there is a correlation between individual's work experience with their working style and organizational performance. Extensive work experience will assist to improve respondent's capabilities in performing PPSM system procedures.

2.2 Comparison Between Frequency of Attending Security Culture Courses and Average Annual Dose Limit

The research identified that the frequencies of attending courses related to security cultures will contribute to the reduction of the average annual dose limit to 71.11% (32 respondents) from 45 respondents who received average annual doses less than 5 mSv and have attended courses related to security cultures more than once as in Table 3.

By frequently attending courses related to security cultures, it will create awareness for the respondents to be more sensitive on the receiving dose, fully complies with the work ethic established by the organization and always practising good work culture. Hence, it will contribute to the reduction of the average annual dose limit received by each respondent.

Table 1 Comparison of the existing PPSM system components based on IAEA documents

No.	Components in PPSM system	Enhancement of PPSM system based on IAEA documents
1.	*Details of radioactive source* i. Types of radionuclide and isotop ii. Activities and measurement date iii. Serial number, physical properties and chemical material iv. Categories of radioactive sources v. Security levels vi. Intended use vii. Supply date of radioactive sources viii. License information	*Details of radioactive source* i. Enhance the inventory system of radioactive sources developed by authority by including categories information for each radioactive sources and security level [18] ii. Medical officers and management team involved should develop an information classification system to identify sensitive information such as details on radioactive source for it to be categorized according to its sensitivity level like secret, confidential and limited [19] iii. Develop a record on device information to store radioactive source information such as model number, devices serial number, manufacturer details and contact number [20]
2.	*Security plan* i. Location ii. Floor plan design iii. Storage and disposal facility iv. Utilization of radioactive source v. Location to install electronic tracking device vi. Security officer vii. Security procedure viii. Security plan objective ix. Procedures to handle with the increased threat level x. Periodically process to evaluate the effectiveness of the securities plan and to update the plan accordingly xi. Reference to the existing regulations or standards	*Security plan* i. Identify and protect computer systems facilities that store, process, manage and transmit sensitive information related to security systems [19] ii. Conduct periodical vulnerability assessment to test and validate whether facilities at medical institution are complied with the security system requirements and to verify the effectiveness of the existing security system [20] iii. Develop a record on personnel/operators who has access to radioactive source area, access key or other relevant system including the computer system [21]
3.	*Security function* Security level i. Security level A ii. Security level B iii. Security level C *Security Function* i. Detect ii. Delay iii. Response iv. Security management	*Security function* i. Security officer in-charge needs to conduct continuous monitoring through the CCTV to ensure immediate detection [11] ii. Provide comprehensive training and relevant tools to the security officer and operators for them to immediately prevent unrestricted access [11] iii. Implement and promote security culture among medical officer involved. Authorities, management team from medical institution, operators and individual are required to involve proactively in identifying and understanding their roles to encourage security culture [22] iv. Authorities need to ensure all medical institution involved are well equipped with security deterrence function to prevent any malicious actions from enemies after assessing the existing security functions that are used for the protection of radioactive sources v. All medical institution involved need to keep radiological emergency response plans as their reference and to ensure emergency situations was handled smoothly

Table 2 Work experience affect understanding on the PPSM procedure

Work experience		Trained and capable of carrying out the procedures in PPSM system			Total
		Yes	No	Not sure	
Less than 5 years	Frequency	6	0	4	10
	Total percentage (%)	11.50	0	7.70	19.20
5–10 years	Frequency	2	2	5	9
	Total percentage (%)	3.80	3.80	9.60	17.30
11–20 years	Frequency	12	9	8	29
	Total percentage (%)	23.10	17.30	15.40	55.80
21–30 years	Frequency	4	0	0	4
	Total percentage (%)	7.70	0	0	7.70
Total	Frequency	24	11	17	52
	Total percentage (%)	46.20	21.20	32.70	100

2.3 *Comparison Between Academic Qualification and Understanding on Basic Concept of PPSM System*

From the research, it shows that understandings on basic concept of PPSM system are not directly affected by respondent's academic qualification with 50% (11 respondents) out of the 22 respondents who have academic qualifications likes certificate/diploma believed detect, inattentive, response and system security management is a basic concept of PPSM system as in Table 4.

Majority of the respondents with academic qualification in bachelor and master's degree are medical officers, physicist officers and biochemical officers. Some of them for example medical officers are basically focused on clinical works whereas physicist and biochemical officers are more involved in administrative activities. As for respondents who possess academic qualification in certificates/diplomas, most of them are radiographer and medical lab technologist which have direct involvement with PPSM system since they are using radioactive source in their daily works.

Table 3 Comparison between frequency of attending security culture courses and average annual dose limit

Attending security culture courses		The average annual dose limit			Total
		Less than 5 mSv	5–10 mSv	11–20 mSv	
Never	Frequency	13	5	2	20
	Total percentage (%)	25	9.60	3.80	38.50
>1	Frequency	29	0	0	29
	Total percentage (%)	55.80	0	0	55.80
>3	Frequency	3	0	0	3
	Total percentage (%)	5.80	0	0	5.80
Total	Frequency	45	5	2	52
	Total percentage (%)	86.50	9.60	3.80	100

Table 4 Comparison between academic qualifications and understanding on basic concept of PPSM system

Academic qualification		The basic concept of PPSM system			Total
		Characterization of facilities, definition of threats, security management system	Detect, delay, response, security management system	Structure, technical, personnel	
Certificate/diploma	Frequency	16	11	3	30
	Total percentage (%)	30.80	21.20	5.80	57.70
Undergraduate	Frequency	9	5	0	14
	Total percentage (%)	17.30	9.60	0	26.90
Postgraduate	Frequency	2	6	0	8
	Total percentage (%)	3.80	11.50	0	15.40
Total	Frequency	27	22	3	52
	Total percentage (%)	51.90	42.30	5.80	100

3 Conclusion

Based on the studies and data analysis conducted, there is a necessity to improve several components in existing PPSM system such as on details components of radioactive source, security plan and functions in order to fulfil current legislative compliance and IAEA recommendations.

Acknowledgements First and foremost, I would like to express my profound gratitude to my advisor, Prof. Dr. Muhamad Samudi bin Yasir for all the guidance and support throughout my research. My sincere thanks also goes to all representatives from MOH, UKMMC, Ampang's Hospital, HUSM, AMDI and Gleaneagles Kuala Lumpur for their participation in this research.

References

1. FOX (2010) 9 years later, nearly 900 9/11 responders have died, survivors fight for compensation. FOX News
2. IAEA (2015b) IAEA welcomes US new global threat reduction initiative. IAEA, Vienna. https://www.iaea.org/newscenter/news/iaea-welcomes-us-new-global-threat-reduction-initiative
3. ElBaradei M (2014) Introductory remarks at the global threat reduction initiative partners conference (GTRI). IAEA, Vienna. https://www.iaea.org/newscenter/statements/introductory-remarks-global-threat-reduction-initiative-partners-conference
4. Amano A (2013) Nuclear security plan 2014–2017, Vienna
5. IAEA (2014) Radiation protection and safety of radiation sources: international basic safety standards. IAEA, Vienna
6. Popp A (2014) Regional security of radioactive sources. ANSTO, New South Wales. http://www.ansto.gov.au/BusinessServices/RegionalSecurityofRadioactiveSourcesProject/
7. Murray A, Popp A, Bus J (2013) Australia's South East Asia regional security of radioactive sources project–achievements and lessons learned. In: International conference on nuclear security: enhancing global efforts. IAEA, Vienna
8. PMO (2010) Nuclear security summit. Office of The Prime Minister, Putrajaya
9. IAEA (2004) Code of conduct on the safety and security of radioactive sources. IAEA, Vienna
10. IAEA (2003) IAEA-TECDOC-1344: categorization of radioactive sources. IAEA, Vienna
11. IAEA (2009a) IAEA nuclear security series no. 11: security of radioactive sources. IAEA, Vienna
12. IAEA (2009b) IAEA nuclear security series no. 10: development, use and maintenance of the design basis threat. IAEA, Vienna
13. Khripunov I (2014) The human dimension of security for radioactive sources. Athens Centre for International Trade and Security, CITS., Badan Tenaga Nuklir Nasional, BATAN
14. Zakariya NIK, Kahn MTE (2015) Safety, security and safeguard. Ann Nucl Energy 75(2015):292–302
15. Gandhi SK, Kang J (2013) Nuclear safety and nuclear security synergy. Ann Nucl Energy 60:357–361
16. Garcia ML (2008) Design and evaluation of physical protection systems. Elsevier Inc., Burlington
17. Garcia ML (2005) Vulnerability assessment of physical protection systems. Elsevier B.V., Burlington
18. IAEA (2005) Safety guide no. RS-G-1.9: categorization of radioactive sources. IAEA, Vienna
19. IAEA (2015a) IAEA nuclear security series no. 23-G: security of nuclear information. IAEA, Vienna

20. IAEA (2016) Revision of nuclear security series no. 11: security of radioactive material in use and storage and of associated facilities. IAEA, Vienna
21. IAEA (2011) IAEA nuclear security series no. 13: nuclear security recommendations on physical protection of nuclear material and nuclear facilities (INFCIRC/225/Revision 5). IAEA, Vienna
22. IAEA (2008) IAEA nuclear security series no. 7: nuclear security culture. IAEA, Vienna
23. Burke LA, Steensma HK (1998) Toward a model for relating executive career experiences and firm performance. J Manag Issues 10(1):86–102

MIX
Papier aus verantwortungsvollen Quellen
Paper from responsible sources
FSC® C105338

If you have any concerns about our products,
you can contact us on
ProductSafety@springernature.com

In case Publisher is established outside the EU,
the EU authorized representative is:
Springer Nature Customer Service Center GmbH
Europaplatz 3, 69115 Heidelberg, Germany

Printed by Libri Plureos GmbH
in Hamburg, Germany